STUDY GUIDE to accompany

Huntsberger, Croft, and Billingsley's

STATISTICAL INFERENCE FOR MANAGEMENT AND ECONOMICS

2nd Edition

Edward R. Mansfield

University of Alabama

ALLYN AND BACON, INC.

Boston, London, Sydney, Toronto

Copyright © 1980 by Allyn and Bacon, Inc.,
470 Atlantic Avenue, Boston, Massachusetts 02210.
No part of the material protected by this
copyright notice may be reproduced or utilized
in any form or by any means, electronic or
mechanical, including photocopying, recording,
or by any information storage and retrieval
system without written permission from the
copyright owner.

ISBN 0-205-06806-5

10 9 8 7 6 5 4 3 2   86 85 84 83 82 81

Printed in the United States of America.

Contents

Review Tests and Problems with Worked Solutions

Chapter 1:	Introduction	1
2:	Empirical Frequency Distributions	6
3:	Descriptive Measures	16
4:	Elementary Probability	27
5:	Populations, Samples, and Distributions	42
6:	Probability Approximations	57
7:	Sampling Distributions	65
8:	Estimation	78
9:	Tests of Hypotheses	100
10:	Analysis of Variance	126
11:	Approximate Tests: Multinomial Data	140
12:	Regression and Correlation	154
13:	Multiple Regression	167
14:	Time Series Analysis	184
15:	Index Numbers	198
16:	Decision Theory	211
17:	Decisions and Experiments	228
18:	Nonparametric Methods	240

Preface

This study guide is designed for use by students who are taking a beginning course in statistics. The questions and solved problems are typical of those covered in most introductory-level courses and are organized to parallel the textbook, <u>Statistical Inference for Management and Economics</u>, by David V. Huntsberger, D. James Croft, and Patrick Billingsley.

Each chapter in this study guide, except the first, consists of three sections designed to aid the student in understanding the material in the text. The first section contains a comprehensive outline of the corresponding text chapter, including key terminology and definitions. This enables the student to obtain an overall view of the objectives of the chapter before reading the text.

The second section of each chapter in this study guide is a review test. This test consists of several multiple choice questions which concentrate on an understanding of the concepts, assumptions, and uses of the material covered in the text. The answers to these questions are given at the end of the Review Test; also given are section references. Students should complete the Review Test before checking any answers. If a student selects an incorrect answer, he or she should reread the appropriate part of the section in the text.

The third part contains several problems comparable to those at the end of the chapters in the text. Many problems are worded in the language of real life or contrived situations, and the student must decide from the nature of the problem which statistical method is needed. At the end of the problem set, a complete solution is given for <u>all</u> problems. These solutions are presented in such a way that the student can follow the logic of the solution. This should enhance the student's ability to approach a problem in a systematic manner.

                                                            ERM

Introduction

Major Topics and Key Concepts

1.1 What is Statistics?

- How does "Statistics" as a science differ from "a statistic," meaning a summary of numbers?

1.2 Types of Problems in Statistics.

- Descriptive statistics.
- Probability.
- Statistical inference.
- Miscellaneous.

1.3 Misuses of Statistics

- Failing to adjust data to a per-item basis.
- Induced bias.
- Inappropriate comparisons of groups such as self-selection and hidden differences.

1.4 Sources of Statistical Data.

- Existing data available.

CHAPTER 1: Review Test

1. Present-day statisticians are mainly concerned with:
    (a) Collecting and tabulating numerical data.
    (b) The development and application of methods and techniques for collecting, analyzing, and interpreting quantitative data.
    (c) The compilation of arrays and tables pertaining to births, deaths, populations, etc.
    (d) All of the above.

2. Business managers and administrators can make use of statistical methods by:
    (a) Being familiar with the theory of probability.
    (b) Employing a statistician as a consultant.
    (c) Using the method of applied statistics.
    (d) All of the above.

3. Descriptive statistics does not deal with:
    (a) Inductive generalizations.
    (b) Organizing and summarizing numerical data.
    (c) The graphical or pictorial presentation of data.
    (d) The calculation of descriptive measures.

4. By making generalizations about a larger group based on the results of a smaller sample taken from that group, one enters into the realm of:
    (a) Descriptive statistics.
    (b) Population statistics.
    (c) Statistical inference.
    (d) Theoretical statistics.

5. When making a decision using statistical inference:
    (a) You are absolutely sure that the correct decision will be made.
    (b) The conclusion is correct only for the items in the sample.
    (c) You have no control over the risk of reaching an incorrect conclusion.
    (d) You realize that there is some chance that an incorrect decision could be made, but you have some control over the probability of that happening.

6. Statistical inference is valid if:
   (a) Samples are taken from very large groups.
   (b) Descriptive statistics cannot be employed.
   (c) Statistical principles are applied in selecting the sample and deriving the estimates.
   (d) Complete information is available.

7. The management of a large corporation (3,000 employees) took a random sample of 100 employees to ascertain their opinion of a new four-day workweek proposal. They found that 55 of the employees sampled favored the four-day plan. Management can correctly assume that the true proportion of all 3,000 employees who favor the four-day work-week is:
   (a) 55%.
   (b) Likely to be greater than 70%.
   (c) Somewhere in the vicinity of 55%.
   (d) No assumptions can be made.

8. In statistics, the theory of chance or probability:
   (a) Can tell whether the sample was large enough.
   (c) Can be used to set reasonable limits on the range in which true proportional values lie.
   (c) Is a vital component of inferences or generalizations made.
   (d) All of the above.

9. Customers in a supermarket are randomly selected and asked to participate in a video-taped taste test. After tasting a new flavored soft drink in front of rolling T.V. cameras, the customers are asked whether or not they would buy this product. By the end of the day, 80% of the respondents claimed they would buy the new product. This result could be misleading because of:
   (a) Failure to adjust data to a per-item basis.
   (b) Induced bias.
   (c) Inappropriate comparisons of groups from self-selection.
   (d) Inappropriate comparisons of groups from hidden differences.

10. A recent federal government study compared the average amount of electricity consumed last summer per household in Texas and in New York state. Since the study showed that a typical Texas household consumed more electricity than a typical New York household, the conclusion reached was that Texans are less willing to conserve energy than New Yorkers. This result could be misleading because of:
    (a) Failure to adjust data to a per-item basis.
    (b) Induced bias.
    (c) Inappropriate comparisons of groups from self-selection.
    (d) Inappropriate comparisons of groups from hidden differences.

11. Which of the following statements is most fitting?
    (a) Mark Twain's statement, "There are three kinds of lies--lies, damned lies, and statistics."
    (b) The comment of a country-western record promoter, "Statistics are great for sales because you can prove anything you want with good statistics."
    (c) The comment of a character in the BC comic strip responding to a question concerning his feelings about surveys, "If you accrued all the intelligence from all the surveys ever taken, and laid them end to end, they would constitute a circle just big enough to encompass a single cow flop."
    (d) The vice-president of product development at Olin Textile Mill's comment, "A basic understanding of statistics is essential for responsible decision making in the face of uncertainty."

12. The broad purpose of the book, <u>Statistical Inference for Management and Economics</u>, is:
    (a) To make statisticians of its readers.
    (b) To help the reader solve complex problems that he would otherwise need the services of a professional statistician to resolve.
    (c) To teach enough about statistical techniques so that the reader may solve many common types of problems himself.
    (d) To imbue the reader with a comprehensive knowledge of all statistical principles and a thorough command of statistical language.

CHAPTER 1: Answers to Review Test

Question	Answer	Text Section Reference
1	b	1.1
2	d	1.1
3	a	1.2
4	c	1.2
5	d	1.2
6	c	1.2
7	c	1.2
8	d	1.2
9	b	1.3
10	d	1.3
11	d	1.2
12	c	1.2

Empirical Frequency Distributions

Major Topics and Key Concepts

2.1 Frequency Distributions

- Definitions of class boundaries, class intervals, class limits, and class marks.
- Construction of frequency distributions.
- Conversion to relative frequency distributions.
- Conversion to percentage distributions.
- A rule for determining number of classes to use.

2.2 Cumulative Frequency Distributions

- How many observations are less than each of the class boundaries.

2.3 Graphic Presentation - The Histogram

- A vertical bar graph of a frequency distribution.

2.4 Graphic Presentation - The Ogive

- A non-decreasing graph of a cumulative frequency distribution.

CHAPTER 2: Review Test

1. In constructing a frequency table one must:
   (a) Divide the overall range of values into a number of mutually exclusive classes.
   (b) Use special frequency table construction paper.
   (c) Use punched-card equipment for large numbers of values.
   (d) Use an array ordering values from low to high.

2. In deciding upon how many classes to use in a frequency distribution, the following applies:
   (a) It is a purely arbitrary decision.
   (b) It is common to use from 5 to 20 classes depending on the size of the data sets.
   (c) Class intervals must be known beforehand.
   (d) All of the above.

3. The use of equal class intervals:
   (a) Is absolutely necessary.
   (b) Gives a clearer picture of lop-sided distributions.
   (c) Results in simpler calculations.
   (d) Is not recommended when working with symmetrical distributions.

4. To prevent ambiguity about which class a given observation belongs to, the person constructing a frequency distribution may:
   (a) Select impossible values for boundaries.
   (b) Use even integers.
   (c) Make use of class marks.
   (d) Set class limits.

5. The representative value of a class is referred to as:
   (a) The class limit.
   (b) The class mark.
   (c) The class boundary.
   (d) The class frequency.

6. The class mark, or midpoint of the class, may be computed by averaging:
   (a) The class limits.
   (b) The class boundaries.
   (c) The class frequencies.
   (d) Both a and b.

7. Relative class frequencies are computed by:
   (a) Dividing the class frequencies by the total number of observations.
   (b) Multiplying class frequencies by 100.
   (c) Multiplying class frequencies by class marks.
   (d) Dividing class marks by the total frequencies for all classes.

8. Open-ended intervals are often useful because:
   (a) The actual numerical value is retained.
   (b) They permit the inclusion of a wide range of extreme values.
   (c) They simplify the calculation of descriptive measures.
   (d) Closed intervals are often too narrow to give meaningful information.

9. Cumulative frequency distributions are valuable when information is required on:
   (a) The number of observations whose numerical value is less than a given value.
   (b) The cumulative number of observations whose numerical value is more than a given value.
   (c) The number of observations that are within some given interval.
   (d) All of the above.

10. A graphic presentation of a frequency distribution that is constructed by erecting vertical bars or rectangles on the class intervals is called:
    (a) A pictograph.
    (b) A histogram.
    (c) An ogive.
    (d) A horizontal bar chart.

11. The class interval used in constructing a frequency table has _____ on the appearance of the corresponding histogram.
    (a) No effect
    (b) A marked effect.
    (c) Little effect.
    (d) Irregular effect.

12. A construction rule that applies to all histograms states that:
    (a) The height of the bar must be proportional to the class frequency.
    (b) The larger the class interval the greater the height of the bar.
    (c) The area of the bar over a class interval must be proportional to the frequency of the class.
    (d) The area of the bar must be proportional to the class interval.

13. Which of the following statements about ogives is not correct?
    (a) An ogive is a graphic representation of a cumulative frequency distribution.
    (b) In constructing an ogive, class boundaries are represented on the vertical scale.
    (c) It is proper to interpolate graphically from an ogive.
    (d) Plotting points are proportional to the cumulative frequencies.

14. Theoretical distributions differ from empirical distributions in that they:
    (a) Cannot be represented graphically.
    (b) Are used to estimate empirical distributions.
    (c) Cannot be used to summarize and extract pertinent information.
    (d) Include the entire population or the totality of possible values which might be obtained.

15. If the class limits in a frequency distribution are 140-143, 144-147, 148-151, and 152-155, what are the boundaries of the classes?
    (a) 140.5-143.5, 144.5-147.5, 148.5-151.5, 152.5-155.5.
    (b) 141.5, 145.5, 149.5, 153.5.
    (c) Same as class limits in this case.
    (d) 139.5-143.5, 143.5-147.5, 147.5-151.5, 151.5-155.5.

16. What are the class marks for the classes given in Problem 15?
    (a) 139.5-143.5, 143.5-147.5, 147.5-151.5, 151.5-155.5.
    (b) 140.5-143.5, 144.5-147.5, 148.5-151.5, 152.5-155.5.
    (c) 141.5, 145.5, 149.5, 153.5.
    (d) 142, 146, 150, 154.

17. The frequencies or numbers of observations for the classes given in Problem 15 are 41, 19, 12, and 2. What are the relative frequencies?

    (a) .5541, .2568, .1622, .0270
    (b) .41, .19, .12, .2
    (c) .6949, .2346, .1364, .0204
    (d) 1, .4634, .2927, .0488

18. If a cumulative distribution were constructed from the information given in Problems 15 and 17, how many of the observations would be less than 151.5?

    (a) 60
    (b) 31
    (c) 72
    (d) 12

19. If the last two classes in Problem 15 were combined and a histogram were constructed, the height of the rightmost bar in the representation would be _____ what it would have been had the classes not been combined.

    (a) one-half
    (b) the same as
    (c) twice
    (d) twelve more than

20. If an ogive were to be constructed from the cumulative distribution developed in Problem 18, the plotting points corresponding to the class boundaries would be:

    (a) 41, 19, 12, 2
    (b) 0, 41, 60, 72, 74
    (c) 0, 41, 60, 31, 14
    (d) 41, 60, 31, 14, 0

CHAPTER 2: Answers to Review Test

Question	Answer	Text Section Reference
1	a	2.1
2	b	2.1
3	c	2.1
4	a	2.1
5	b	2.1
6	d	2.1
7	a	2.1
8	b	2.1
9	d	2.2
10	b	2.3
11	b	2.3
12	c	2.3
13	b	2.4
14	d	2.4
15	d	2.1
16	c	2.1
17	a	2.1
18	c	2.2
19	a	2.3
20	b	2.4

CHAPTER 2: Review Problems

1. The following scores were obtained on the final examination in an elementary statistics course:

60	79	48	57	74	58	70	82	77	81	95	100
65	92	85	55	52	64	75	78	55	80	98	81
71	83	54	84	72	88	62	74	78	89	76	84
48	84	90	85	67	52	82	69	74	73	80	95

    (a) Construct the frequency distribution.
    (b) Give the class boundaries and class marks.
    (c) Construct the cumulative distribution and cumulative relative frequency distribution.
    (d) Draw the histogram and ogive.

2. Given this histogram for times needed to repair a tape recorder:

    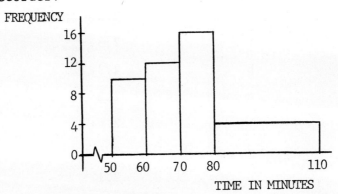

    (a) Determine the sample size.
    (b) Estimate the number of times that were less than 65 minutes.

12

CHAPTER 2: Solutions to Review Problems

1. <u>Solution</u>

   a,b First, decide on the number of classes needed and the width of each class. Since n = 48, the first power of 2 which equals or exceeds 48 is 6 (2^6 = 64); hence, 6 classes should be used. The range (100 - 48 = 52) divided by the number of desired classes (52/6 = 8.667) indicates that each class should have an interval of 9. Choosing 48 to 57 as the first class, the remaining classes can be defined. Use a tally sheet to count frequencies.

FINAL EXAM SCORE	FREQUENCY	BOUNDARIES	CLASS MARKS
48 to 56	7	47.5	52
57 to 65	6	56.5	61
66 to 74	9	65.5	70
75 to 83	13	74.5	79
84 to 92	9	83.5	88
93 to 101	4	92.5	97
	48	101.5	

   Note that the sum of the frequencies must equal the sample size.

   c When constructing the cumulative frequency distribution, always use the class boundaries. Note that the last cumulative frequency must equal the sample size.

FINAL EXAM SCORE	CUMULATIVE FREQUENCY	CUMULATIVE RELATIVE FREQUENCY
Less than 47.5	0	0
Less than 56.5	7	.1458
Less than 65.5	13	.2708
Less than 74.5	22	.4583
Less than 83.5	35	.7292
Less than 92.5	44	.9167
Less than 101.5	48	1.0000

   The cumulative relative frequencies are

obtained by dividing each cumulative frequency by the sample size, 48.

d   When constructing any graph or table, always be sure to label the axes. The vertical axis must be able to accommodate the largest frequency for a histogram and the sample size for an ogive.

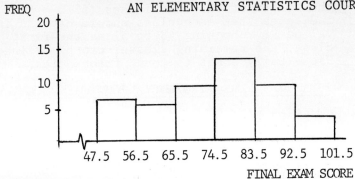

HISTOGRAM FOR THE FINAL EXAMINATION IN AN ELEMENTARY STATISTICS COURSE

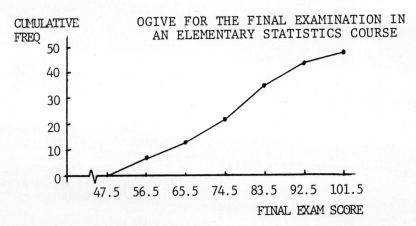

OGIVE FOR THE FINAL EXAMINATION IN AN ELEMENTARY STATISTICS COURSE

2. <u>Solution</u>

a   The frequencies for the four classes are 10, 12, 16, and 12; hence, n = 50. Note that the interval of the 80 to 110 class is three times larger than the other classes; hence, its frequency is 4 · 3 = 12.

b   Construction of an ogive facilitates the interpolation.

14

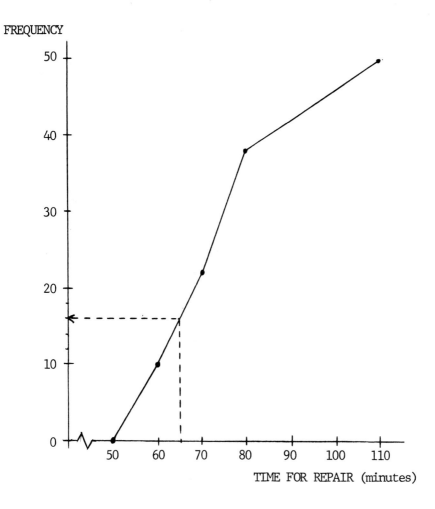

Therefore, the number of times that were less than 65 minutes is 16.

Descriptive Measures

Major Topics and Key Concepts

3.1 Symbols and Summation Notation

3.2 Measures of Location (or Central Tendency)

- The mean, median, and mode.
- For raw data, see sections 3.3, 3.4, 3.5, 3.6; for grouped data or frequency distributions, see section 3.13.

3.3 The Mean

- The arithmetic average of n values.
- The symbol μ denotes the population mean (a parameter) and can be calculated only when the entire population can be enumerated.
- The symbol $\bar{X}$ denotes the sample mean (a statistic) and is an estimate of the parameter μ.

3.4 The Weighted Mean

- Used when some observed values carry more importance or higher weight than other observations.

3.5 The Median

- The value that falls in the middle of the ordered set of data.

3.6 The Mode

- The value or values that occur the most frequently.

3.7 Measure of Variation

- For raw data, see sections 3.7 through 3.11; for grouped data, see section 3.13.

3.8 The Range

- The distance between the largest and smallest values in the data.

3.9 Deviations from the Mean

- The sum of the absolute distances from the mean divided by n.

3.10 The Variance and Standard Deviation

3.11 The Sample Standard Deviation

- Equations

	For the Entire Population	For a Sample
Variance	$\sigma^2 = \dfrac{\Sigma(X_i - \mu)^2}{n}$	$s^2 = \dfrac{\Sigma(X_i - \bar{X})^2}{n-1}$ or $s^2 = \dfrac{\Sigma X_i^2 - \dfrac{(\Sigma X)^2}{n}}{n-1}$
Standard Deviation	$\sigma = \sqrt{\dfrac{\Sigma(X_i - \mu)^2}{n}}$	$s = \sqrt{s^2}$

3.12 What Does the Standard Deviation Tell Us?

3.13 Descriptive Statistics for Grouped Data

- Mean, median, mode, and variance.

3.14 Selecting a Measure of Location

3.15 The Coefficient of Variation

3.16 Descriptive Measures and Electronic Computers

CHAPTER 3:   Review Test

1. All of the following are measures of central tendency or location, except

    (a) The mean.
    (b) The median.
    (c) The range.
    (d) The mode.

2. For a given set of n values, X_1, X_2, X_3, ..., X_n, the mean is

    (a) The most frequently occurring value.
    (b) The difference between the largest and smallest value.
    (c) Their sum divided by n-1 (degrees of freedom).
    (d) Their sum divided by n.

3. The weighted mean is found by dividing the sum of the products of the values and their weights by

    (a) The sum of the weights.
    (b) The number of values.
    (c) The total number of values minus one.
    (d) The sum of the values.

4. Which of the following statements concerning the mode is correct?

    (a) The mode is a measure of variation.
    (b) The mode is frequently used as a measure of location for ungrouped data.
    (c) The mode is the most frequently occurring value in the data set.
    (d) All of the above.

5. The range differs from most of the statistical measures in that it is a relatively good measure of variation for

    (a) Comparing two sets of data with widely varying numbers of values.
    (b) Small number of observations.
    (c) Large samples.
    (d) Use in statistical quality control.

6. Which of the following statements about the standard deviation is correct?

    (a) It is the square root of the variance.
    (b) Its units are the same as those of the original data.
    (c) It cannot be calculated for grouped data.
    (d) Both (a) and (b).

7. The major difference(s) between σ^2 and s^2 is (are):
   (a) σ^2 pertains to the entire population where s^2 is calculated from a sample and is used to estimate σ^2.
   (b) σ^2 has n in the denominator where s^2 has n-1.
   (c) σ^2 is a parameter whereas s^2 is a statistic.
   (d) All of the above are true.

8. Suppose that IQ scores for a particular population follow a bell-shaped pattern and the true population mean for all the IQ scores is $\mu = 108$ and true population standard deviation is $\sigma = 14$. Then about two-thirds of all the individual IQ scores should fall between:
   (a) 94 and 122
   (b) 101 and 115
   (c) 80 and 136
   (d) 66 and 150

9. An expression defined as the number of squares minus the number of independent linear restrictions imposed upon the quantities involved is known as its
   (a) Variance.
   (b) Weighted mean.
   (c) Degrees of freedom.
   (d) Standard deviation.

10. For grouped data, the mean cannot be evaluated accurately, if at all, when:
    (a) The frequency distribution includes open-ended classes.
    (b) The frequency distribution is unimodal.
    (c) Class intervals have equal widths.
    (d) The dispersion is great.

11. In general, the wider a class interval, the
    (a) Less loss of information through grouping.
    (b) Farther the grouped mean tends to be from the actual, ungrouped mean.
    (c) Closer the grouped mean is to the ungrouped mean.
    (d) Easier it is to determine the class mark.

12. In estimating the median for grouped data by use of the arithmetic procedure, which of the following pieces of information is needed?
    (a) Lower boundary of the class containing the median.
    (b) Frequency of the class containing the median.
    (c) Length of the class containing the median.
    (d) All of the above.

13. If the distribution of the data is symmetric and unimodal, then the mean, the median, and the mode:

    (a) Will all coincide.
    (b) Will not be determinable.
    (c) May be misleading.
    (d) May all be to the right of center.

14. The _____ is more stable, more amenable to mathematical and theoretical treatment, and an almost universal choice for all but purely descriptive purposes.

    (a) Mode
    (b) Median
    (c) Mean
    (d) Range

15. Which of the following measures of location is most sensitive to extreme values?

    (a) The mode.
    (b) The mean.
    (c) The median.
    (d) All are equally sensitive.

16. When data are skewed, the best description of central tendency is most likely to be provided by:

    (a) The median.
    (b) The mean.
    (c) The mode.
    (d) The median or the mode.

17. When calculated from grouped data, the variance and standard deviation are, in general:

    (a) Not exact due to information lost in grouping.
    (b) Always good approximations.
    (c) Good approximations if class intervals are sufficiently large.
    (d) As exact as they would be for ungrouped data.

18. The coefficient of variation is a useful measure for comparing the relative dispersion of two or more samples when:

    (a) The samples are measuring unrelated variables.
    (b) The magnitudes of the data in the samples are different.
    (c) The sample means are all equal.
    (d) One of the variances is identical to the user's age at his last birthday.

19. Given that $X_1 = 5$, $X_2 = 7$, $X_3 = 4$, $X_4 = 8$, $X_5 = 6$, then ΣX^2 and $(\Sigma X)^2$, respectively, are:
    (a) 672 and 163
    (b) 900 and 190
    (c) 190 and 900
    (d) 163 and 672

20. The mean of the data given in Problem 19 is:
    (a) 4
    (b) 6
    (c) 30
    (d) 7.5

21. The sample variance and sample standard deviation for the data in Problem 19 are:
    (a) 180 and 13.416
    (b) 10 and 3.162
    (c) 38 and 6.164
    (d) 2.5 and 1.581

(Use the following table for Problems 22 through 25.)

The following table is the frequency distribution of the percentage return on investment in 1973 for 30 business firms:

Percentage Return on Investment	(Frequency) Number of Companies
6.00 but less than 6.50	5
6.50 but less than 7.00	8
7.00 but less than 7.50	10
7.50 but less than 8.00	7

22. The mean net return on investment is:
    (a) 7.25%
    (b) 7.00%
    (c) 6.99%
    (d) 7.07%

23. The median rate of return for this data is:
    (a) 7.10%
    (b) 7.20%
    (c) 7.25%
    (d) 7.00%

24. The sample standard deviation for return data is:
    (a) .5167
    (b) .6455
    (c) 2.58
    (d) 16.89

25. The degrees of freedom are:
    (a) 3
    (b) 4
    (c) 29
    (d) 30

CHAPTER 3: Answers to Review Test

Question	Answer	Text Section Reference
1	c	3.2
2	d	3.3
3	a	3.4
4	c	3.6
5	b and d	3.8
6	d	3.11
7	d	3.10 and 3.11
8	a	3.12
9	c	3.11
10	a	3.13
11	b	3.13
12	d	3.13
13	a	3.14
14	c	3.14
15	b	3.14
16	a	3.14
17	a	3.13
18	b	3.15
19	b	3.1
20	b	3.2
21	d	3.11
22	d	3.13
23	a	3.13
24	a	3.13
25	c	3.11 and 3.13

CHAPTER 3:   Review Problems

1. The number of touchdowns scored in each game during one season by a team contending for a national championship was:

   $$2, 8, 2, 5, 0, 6, 6, 3, 6, 4, 2$$

   (a) Find the mean, median, and mode number of touchdowns per game.
   (b) Find the range.
   (c) Find the standard deviation.

2. Six individuals were used in a controlled study to determine if disco dancing is an effective means of losing weight. None of the six individuals were disco dancers before the study; all were moderately overweight. The weight losses (before weight - after weight) were:

   $$2, -15, 3, 4, -5, 6$$

   (a) Find the mean and median weight loss.
   (b) Find the standard deviation.
   (c) Does this data indicate that disco dancing is effective for losing weight?

3. The following table represents the distance in centimeters from the center of the target to a blow dart for 40 shots by a male Aborigine warrior:

Distance (centimeters)	Frequency
0.0 but less than 4.0	3
4.0 but less than 8.0	1
8.0 but less than 12.0	4
12.0 but less than 16.0	6
16.0 but less than 20.0	0
20.0 but less than 24.0	10
24.0 but less than 28.0	10
28.0 but less than 32.0	6

   (a) Find the mean and median distance from center.
   (b) Calculate the variance and standard deviation
   (c) Into which class would you say that the "typical" shot would fall? Why?

4. Find the mean, median, and modal class:

Class	Relative Frequency
4 to 7	.10
8 to 11	.30
12 to 15	.40
16 to 19	.20

CHAPTER 3: Solutions to Review Problems

1. <u>Solution</u>

   a. The <u>mean</u> is $\bar{X} = \frac{\Sigma X}{n}$,

   which is $\bar{X} = (2 + 8 + 2 + \ldots + 2)/11 = 44/11 = 4$.

   The <u>median</u> is the observation that falls in the middle of the <u>ordered</u> sample. Arrange the data in ascending order:

   $$0, 2, 2, 2, 3, 4, 5, 6, 6, 6, 8.$$

   Since there are 11 observations, the median is the 6th value; i.e., median = 4.

   The <u>mode</u> is the value which occurs the most frequently. For this data, both 2 and 6 appear 3 times. Therefore, the modes are 2 and 6.

   b. The range = 8 − 0 = 8.

   c. The variance can be found by using either formula. We will use $s^2 = \frac{\Sigma(X_i - \bar{X})^2}{n-1}$. $\bar{X} = 4$ and $n = 11$.

X_i	$X_i - \bar{X}$	$(X_i - \bar{X})^2$
2	−2	4
8	4	16
⋮	⋮	⋮
4	0	0
2	−2	4
44	0	58

   $s^2 = \frac{58}{10} = 5.8$

   Hence, the standard deviation is

   $s = \sqrt{5.8} = 2.408$.

2. <u>Solution</u>

   a. $\bar{X} = \frac{\Sigma X}{n} = \frac{-5}{6} = -.833$ pounds.

   The mean indicates a small average weight gain. Since the sample is 6 (even), the <u>median</u> will be average of the middle two values in the <u>ordered</u> sample: −15, −5, 2, 3, 4, 6.

   Hence, the median = $\frac{2 + 3}{2}$ = 2.5 pounds lost.

   b. Using the machine formula

   $s^2 = \frac{\Sigma X^2 - \frac{(\Sigma X)^2}{n}}{n-1}$.

$\Sigma X = -5$, $\Sigma X^2 = 2^2 + (-15)^2 + 3^2 + 4^2 + (-5)^2 + 6^2 = 315$.

Hence, $s^2 = \dfrac{315 - \dfrac{(-5)^2}{6}}{5} = \dfrac{315 - 104.1667}{5} =$

$= 62.1667$ and $s = 7.885$.

c. The one extreme weight <u>gain</u> of 15 pounds was probably not due to disco dancing. Using the median as the measure of location, there is some indication of a small weight loss.

3. <u>Solution</u>

a,b. The mean for grouped data is $\bar{X} = \dfrac{\Sigma f_i v_i}{n}$

and the variance is $s^2 = \dfrac{\Sigma f_i v_i^2 - \dfrac{(\Sigma f_i v_i)^2}{n}}{n-1}$

We will do these parts together:

Distance (centimeters)	f_i	Marks v_i	$f_i v_i$	v_i^2	$f_i v_i^2$
0.0 but less than 4.0	3	2.0	6	4	12
4.0 but less than 8.0	1	6.0	6	36	36
8.0 but less than 12.0	4	10.0	40	100	400
12.0 but less than 16.0	6	14.0	84	196	1176
16.0 but less than 20.0	0	18.0	0	324	0
20.0 but less than 24.0	10	22.0	220	484	4840
24.0 but less than 28.0	10	26.0	260	676	6760
28.0 but less than 32.0	6	30.0	180	900	5400
	40		796		18624

Therefore, $\bar{X} = \dfrac{796}{40} = 19.9$

$s^2 = \dfrac{18624 - \dfrac{(796)^2}{40}}{39} = 71.374$

and $s = \sqrt{71.374} = 8.448$

The <u>median</u> would be the 20th closest shot. This would fall in the class, "20 but less than 24." Since 14 shots occur before this class, $d = 20 - 14 = 6$ more shots must be counted into this class. The lower class boundary is 20.0. Therefore,

$$\text{median} = b + \frac{d}{f} \cdot c = 20.0 + \left(\frac{6}{10} \cdot 4.0\right) = 22.4.$$

c. Because the distribution is skewed, use the median as measure of location.

4. <u>Solution</u>

Even without the sample size, we can still find these statistics;

Class	Relative Frequencies r_i	v_i	$r_i \cdot v_i$
4 to 8	.10	6.0	.6
9 to 13	.30	11.0	3.3
14 to 18	.40	16.0	6.4
19 to 28	.20	23.5	4.7
			15.0

To find the mean,

since $r_i = \frac{f_i}{n}$, $\bar{X} = \frac{\Sigma f_i v_i}{n} = \Sigma r_i v_i = 15.0$.

The median value occurs when the relative frequency is 0.5. The median is in the class "14 to 18" which has a lower class boundary of 13.5 and interval of 18.5 - 13.5 = 5.0. Here, d = .50 - (.10 + .30) = .10; therefore,

$$\text{median} = 13.5 + \frac{.10}{.40} \cdot 5.0 = 14.75 .$$

4

Elementary Probability

Major Topics and Key Concepts

4.1 The Meaning of Probability

- Sample space, outcomes, events.
- Types of probabilities: classical, relative frequency, subjective.

4.2 The Addition Principles

- Mutually exclusive events: no outcomes in common.
- Determining the number of objects in either or both of two sets.

4.3 The Multiplication Principle

- Counting the total number of possible choices that can be made when chosing at each of two or more independent stages.

4.4 Permutations and Combinations

- Factorial notation: $n! = n(n-1) \cdot \ldots \cdot 3 \cdot 2 \cdot 1$.
- Combinatorial notation: $\binom{n}{r} = \frac{n!}{r!(n-r)!}$ (see Table I)
- Counting the number of ways in which r items can be selected from n items when:
    1. order of selection matters - permutations; $\frac{n!}{(n-r)!}$
    2. order of selection does not matter - combinations; $\binom{n}{r}$.

4.5 Computing Probabilities

- Addition of Probabilities.
- Complement of an event.
- Conditional probability.
- Independence.

SUMMARY	P(A and B) =	P(A or B) =
If A and B are Mutually Exclusive Events	0	P(A) + P(B)
If A and B are Independent Events	P(A) · P(B)	P(A) + P(B) − P(A) · P(B)
General Case	P(A) · P(B|A)	P(A) + P(B) − P(A and B)

4.6 Probability Tree Diagrams

4.7 Subjective Probability and Bayes' Theorem

$$P(A \text{ and } B) = P(A) \cdot P(B/A)$$

$$P(B/A) = \frac{P(A)}{P(A \text{ and } B)}$$

CHAPTER 4: Review Test

1. The set of all possible outcomes or results of an experiment is called the
   - (a) Event.
   - (b) Favorable cases.
   - (c) Sample space.
   - (d) Subset.

2. The probability of an event is taken to be:
   - (a) The ratio of the number of cases favorable to the event to the number of cases in the sample space.
   - (b) The ratio of total number of cases to number of favorable cases.
   - (c) The sample space minus the favorable cases.
   - (d) The ratio of favorable cases to unfavorable cases.

3. Upon arrival in Las Vegas, a young Texan, unfamiliar with the game of Blackjack, had the rules explained to him for the first time. After some thought at the bar, he determined that his probability of beating the dealer was greater than 0.5. The type of probability that he found was:
   - (a) Classical.
   - (b) Relative frequency.
   - (c) Subjective.

4. Determined to learn more about the game, the Texan observed several thousand hands of Blackjack and noted that the dealer won 54% of the time. This type of probability is
   - (a) Classical.
   - (b) Relative frequency.
   - (c) Subjective.

5. When comparing two sets of objects, if there is no object that belongs to both sets, then the two sets are said to be:
   - (a) Subsets.
   - (b) Disjoint.
   - (c) Mutually independent.
   - (d) Independent.

6. In the General Addition principle for sets (a + b - c) the "c" represents:
   - (a) Zero, if the sets being added are disjoint.
   - (b) An error correction factor.
   - (c) The number of objects found in both sets.
   - (d) (a) and (c).

29

7. Referring to Problem 6, if two events A and B are mutually exclusive or disjoint, then:

    (a) "c" = 0.
    (b) "c" > 0.
    (c) Either "a" or "b" is equal to zero.
    (d) "a" + "b" < "c".

8. One of the conditions that must be satisfied if the multiplication principle is to be applicable is:

    (a) There can be no more than three alternative stages or sequences.
    (b) First-stage alternatives coupled with second-stage successors must lead to distinct results.
    (c) The different choices must lead to similar results.

9. It is sometimes difficult to determine whether a problem requires combinations or permutations. The key is to decide if the ordering of the objects matters:

    (a) If arrangement or order is irrelevant, permutations are called for.
    (b) If order is relevant, combinations are called for.
    (c) If arrangement or order is relevant, permutations are called for.
    (d) If there are n items to be taken r at a time, combinations are called for.

10. If 50 candidates are competing for the Miss America crown, in how many ways can the winner and four runners-up be selected?

    (a) The order of selection is relevant; therefore, the number is $\frac{50!}{45!} = 254,251,200$.
    (b) The order of selection is relevant; therefore, the number is $\frac{50!}{5!} = 2.5345 \times 10^{62}$.
    (c) The order of selection does not matter; therefore, the number is $\binom{50}{5} = 2,118,760$.
    (d) The order of selection does not matter, therefore, the number is $\frac{50}{5} = 45$.

11. If 50 high school seniors are competing for five college scholarships, then how many ways can the winners be selected?

    (a) The order of selection is relevant; therefore, the number is $\frac{50!}{45!} = 254,251,200$.

(b) The order of selection is relevant; therefore, the number is $\frac{50!}{5!} = 2.5345 \times 10^{62}$.

(c) The order of selection does not matter; therefore, the number is $\binom{50}{5} = 2,118,760$.

(d) The order of selection does not matter; therefore, the number is $\frac{50}{5} = 45$.

12. Probabilities exhibit the following general property (properties):
    (a) $0 \leq P(A) \leq 1$.
    (b) $P(A \text{ or } B) = P(A) + P(B) - P(A \text{ and } B)$.
    (c) $P(A \text{ and } B) = P(A) \; P(B|A)$.
    (d) All of the above.

13. When two events are independent, this means that:
    (a) The occurrence or non-occurrence of one event will not influence the probability of the occurrence or non-occurrence of the other event.
    (b) The probabilities of their occurrences must be equal.
    (c) If one of the events does occur, then the other will not occur.
    (d) The probability that both events occur is equal to one minus the product of the individual probabilities.

14. If two events with non-zero probability are known to be independent, then:
    (a) They must also be mutually exclusive.
    (b) They may be mutually exclusive, but not necessarily.
    (c) They can never be mutually exclusive.
    (d) There is insufficient information to determine whether or not the events are mutually exclusive.

15. Events A and B are called independent if:
    (a) $P(A \text{ and } B) = P(A) + P(B)$.
    (b) $P(B|A) = P(A)$.
    (c) $P(A \text{ and } B) = P(A) \cdot P(B)$.
    (d) All of the above.

16. Which of the following is not a characteristic of tree diagrams?
    (a) The probabilities at the end of the branches are joint probabilities which all sum to one.
    (b) The probabilities associated with all the branches after the first set are conditional probabilities.

- (c) All the branches emanating from one point have probabilities that sum to one.
- (d) Each event represented on the diagram is independent of all other events on the diagram.

17. In the conception of subjective probability:
   - (a) The probability of an event is considered a property of the event itself.
   - (b) Numerical probabilities are assigned according to degrees of belief.
   - (c) Repeated trials are required.
   - (d) Relative frequencies tend to stabilize.

CHAPTER 4: Answers to Review Test

Question	Answer	Text Section Reference
1	c	4.1
2	a	4.1
3	c	4.1
4	b	4.1
5	b	4.2
6	d	4.2
7	a	4.2
8	b	4.3
9	c	4.4
10	a	4.4
11	c	4.4
12	d	4.5 and Summary
13	a	4.5 and Summary
14	c	4.5 and Summary
15	c	4.5 and Summary
16	d	4.6
17	b	4.7

CHAPTER 4: Review Problems

1. Use the sample space that occurs when rolling two dice to answer the following questions.

    Let A be the event of rolling a 6 on either die.
    Let B be the event of rolling a total of 6.
    Let C be the event of rolling a total of 9.

    (a) Find P(A), P(B), and P(C).
    (b) Find P(A or B), P(A or C).
    (c) Are A and B mutually exclusive? Are A and C?
    (d) Are A and C independent?

2. In an ordinary deck of 52 cards, how many of the cards are either RED or FACE CARDS?

3. How many different five-card poker hands could be dealt?

4. How many different ways can 12 finalists in the Olympic figure skating competition be ranked?

5. (a) How many different five-man starting teams can be selected from a basketball squad consisting of 13 men?

    (b) If Hearne, the all-star center, and Beckett, the team captain, must be starters, then how many different starting teams can be selected?

    (c) If an 11-man travel team is to be selected to tour Japan, how many possible teams could result from the 13-man roster? Use Table I.

6. A department store has a chief window dresser, Stephanie, charged with decorating all the windows around the store.

    (a) The north side of the store has four display windows. The window dresser has decided that she will put one dress from each of the store's four chief lines in each of these four windows. A "W" line dress will go in the first window, an "X" line dress will go in the second window, a "Y" line dress in the third, and a "Z" line dress in the fourth. How many different groups of four dresses could she pick if she has 10 line "W" dresses, 22 line "X" dresses, 6 line "Y" dresses, and 5 line "Z" dresses?

    (b) Assume Stephanie has now moved on to the five display windows along the east side of the store. She wants to display furniture in these windows. She has narrowed her choice down to 8 sets of

34

furniture. She has decided that she will select 5 sets of furniture at a time and put them in the windows. How many possible orderings of 8 sets in 5 windows can she make?

(c) On the west side of the store the window dresser wants to display men's suits. She has seven different suits she could display. She has three windows in which she can display them. She is not concerned with the order of the suits in the windows. How many different displays can she come up with?

7. One hundred people who have heard about a certain product were interviewed for their opinions. Forty people said they thought they would buy the product in the near future. Sixty people said they didn't think they would buy it. Eighty people said they had used the product in the past, and twenty said they had not used it. Twenty-five of the people who had used the product said they planned to buy it in the near future.

   (a) Find the probability that someone will buy the product if they have tried it (based on these results).
   (b) Find the probability that someone will not buy the product again once they have tried it.
   (c) Given that someone says they will buy the product in the near future, what is the probability that they have used it before?

8. Let A be the event that the temperature drops below 32°F (0°C) and let B be the event that is snows. Given that $P(A) = .75$, $P(B) = .32$, and $P(A \text{ and } B) = .28$,

   (a) Find the probability of either freezing temperature or snow.
   (b) Find the probability that it freezes given that it snows.
   (c) Are A and B independent?
   (d) Are A and B mutually exclusive?

9. The Rho Beta Blues are playing for the intramural basketball championship. There is no time left on the clock, but the Blues center, Dave Dickson, is fouled at the buzzer in the act of shooting. The score is Rho Beta 84, Delta Downers 85. Past records show that Dickson makes 80% of his first foul shots when he has two shots coming to him. But if he misses the first shot, the records show that he makes only 40% of his second shots. If he makes the first shot, there is a 90% chance that he will make the second. Given that

he was fouled while shooting and will get two free throws,

   (a) What is the probability that Rho Beta will lose the game by a score of 85 to 84?
   (b) What is the probability that there will be an overtime played due to a tie score of 85 to 85 at the end of regulation play?
   (c) What is the probability that Rho Beta will win 86 to 85?

10. The probability of selling a home during an open house depends upon weather. If it rains, the probability is only .04; whereas, with no rain, the probability is .20. Suppose the expected probability for rain next Sunday is .30, what is the probability that the home sells?

11. The Penquin Company employs 200 men and 50 women. Of the male employees, 140 work in the plant, 20 in the office, and 40 are field salesmen. The female employees are distributed as follows: 10 to the plant, 25 to the office, and 15 to sales. If an employee is selected at random, what is the probability that he or she is:

   (a) Male?
   (b) Female?
   (c) A plant employee?
   (d) An office employee?
   (e) A sales person?
   (f) A male plant employee?
   (g) A female office employee?
   (h) Plant worker given that he is male?
   (i) An office employee given that she is female?

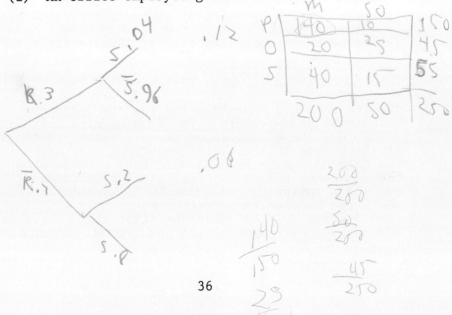

CHAPTER 4: Solutions to Review Problems

1. <u>Solution</u>

   Denote the sample space with the following figure.

   a. Outcomes favorable to events A, B, and C are marked with the appropriate letter. Hence

   $P(A) = \frac{11}{36}$, $P(B) = \frac{5}{36}$

   $P(C) = \frac{4}{36}$.

	1	2	3	4	5	6
1					B	A
2					B	A
3				B		C A
4			B		C	A
5	B				C	A
6	A	A	C A	A	A	A

   b. $P(A \text{ or } B) = \frac{11+5}{36} = \frac{16}{36}$.

   $P(A \text{ or } C) = \frac{11+4-2}{36} = \frac{13}{36}$.

   c. Since A and B have no outcomes in common, they are mutually exclusive; however, A and C have 2 points in common and are not mutually exclusive.

   d. Since P(A and C) = 2/36 is not equal to

   $P(A) \cdot P(C) = \frac{44}{1296}$, these events are not independent.

2. <u>Solution</u>

   Using the General Addition Principle:

   26 RED CARDS + 12 FACE CARDS - 6 RED FACE CARDS

   = 32 cards .

3. <u>Solution</u>

   The order of selection is irrelevant; the number of possible hands is

   $_{52}C_5 = \binom{52}{5} = \frac{52!}{5! \, 47!} = \frac{52 \cdot 51 \cdot 50 \cdot 49 \cdot 48}{5 \cdot 4 \cdot 3 \cdot 2 \cdot 1}$

   = 2,598,960 .

4. <u>Solution</u>

   The ranking of the skaters is important; therefore we want permutations:

   $_{12}P_{12} = \frac{12!}{0!} = 479,001,600$ different possible rankings.

5. Solution

   a. Since the order is not pertinent,

   $$_{13}C_5 = \binom{13}{5} = \frac{13 \cdot 12 \cdot 11 \cdot 10 \cdot 9}{5 \cdot 4 \cdot 3 \cdot 2 \cdot 1} = 1287 \text{ possible starting teams.}$$

   (Note that you can simplify before multiplying.)
   Using Table I in the text, enter with n = 13, r = 5 and read 1287.

   b. If two slots are already filled, then the remaining 3 positions must be chosen from the other 11 players; hence,

   $$_{11}C_3 = \binom{11}{3} = \frac{11!}{3! \; 8!} = 165.$$

   Confirm using Table I.

   c. Again, order does not matter; we want $_{13}C_{11}$.
   Using Table I, n = 13, r = 11, we see that r = 11 is not given. Observe the footnote in Table I.

   $$\binom{13}{11} = \binom{13}{13-11} = \binom{13}{2} = 78 \text{ different possible travel teams.}$$

6. Solution

   a. On the north side each display is filled independently; hence, using the multiplication rule (Section 4.3),

   window: 1   2   3   4
   # of choices = 10 · 22 · 6 · 5 = 6600 different groups.

   b. On the east side, she is concerned about how the 5 chosen furniture sets are ordered (known to us only because she says so). Therefore,

   $$_8P_5 = \frac{8!}{3!} = 8 \cdot 7 \cdot 6 \cdot 5 \cdot 4 = 6720 \text{ permutations.}$$

   c. On the west side, 7 suits chosen 3 at a time is

   $$_7C_3 = \binom{7}{3} = 35 \text{ combinations.}$$

7. Solution

   Two questions are asked in the survey: Would you buy the product in the near future? Have you used the product in the past? This is handled best by forming a table:

		HAVE USED PRODUCT?		
		Yes	No	
WILL BUY PRODUCT IN FUTURE?	Yes	25	15	40
	No	55	5	60
		80	20	100

    a.   Pr (Will Buy | Tried It) = 25/80 = .3125.

    b.   Pr (Will Not Buy | Tried It) = 55/80 = .6875.

    c.   Pr (Have Used It | Will Buy It) = 25/40 = .6250.

8. Solution

    a. $P(A \text{ or } B) = P(A) + P(B) - P(A \text{ and } B) =$
         $.75 + .32 - .28 = .79$.

    b. $P(A|B) = \dfrac{P(A \text{ and } B)}{P(B)} = \dfrac{.28}{.32} = .875$.

    c.   Since $P(A|B) \neq P(A)$ or since $P(A) \cdot P(B) \neq P(A \text{ and } B)$, these two events are not independent.

    d.   Since $P(A \text{ and } B) \neq 0$, they are not mutually exclusive.

9. Solution

Let H_1 be the event of a hit on the first shot and H_2 be the event of a hit on the second shot. We are given $P(H_1) = .80$, $P(H_2|\overline{H}_1) = .40$, $P(H_2|H_1) = .90$.

    a.   Rho Beta losing means Dickson misses both shots:
         $P(\overline{H}_1 \text{ and } \overline{H}_2) = P(\overline{H}_1) \cdot P(\overline{H}_2|\overline{H}_1) = (.20) \cdot (.60) = .12$.

    b.   Overtime means exactly one hit on either the first or the second shot:
         $P(\text{Overtime}) = P(\overline{H}_1 \text{ and } H_2) + P(H_1 \text{ and } \overline{H}_2)$
                     $= P(\overline{H}_1) \cdot P(H_2|\overline{H}_1) + P(H_1) \cdot P(\overline{H}_2|H_1)$
                     $= (.20) \cdot (.40) + (.80) \cdot (.10) = .16$.

c. $P(WIN) = P(H_1 \text{ and } H_2) = P(H_1) \cdot P(H_2|H_1) = (.8)(.9) = .72$.

NOTE: $P(LOSE) + P(OVERTIME) + P(WIN) = 1$.

a, b and c Using a tree diagram:

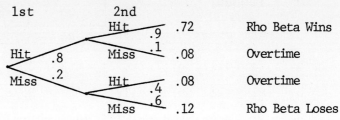

10. **Solution**

    $P(\text{Home Sells}) = P(\text{Sells and Rain}) + P(\text{Sells and Does Not Rain})$

    $= P(\text{Rain}) \cdot P(\text{Sells}|\text{Rain}) + P(\text{Does Not Rain}) \cdot P(\text{Sells}|\text{No Rain})$

    $= (.3) \cdot (.04) + (.7) \cdot (.20)$

    $= .012 + .14 = .152$.

    Using a tree diagram

    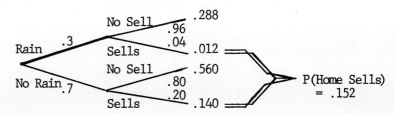

11. **Solution**

    The problem can more readily be solved by putting the information in tabular form.

Location	Male (M)	Female (F)	Totals
Plant (P)	140	10	150
Office (O)	20	25	45
Sales (S)	40	15	55
Totals	200	50	250

(a) $P(\text{Male}) = \dfrac{200}{250} = .80$ \qquad (b) $P(\text{Female}) = \dfrac{50}{250} = .20$

(c) $P(P) = \dfrac{150}{250} = .60$ \qquad (d) $P(O) = \dfrac{45}{250} = .18$

(e) $P(S) = \dfrac{55}{250} = .22$

(f) $P(M \text{ and } P) = P(M) \cdot P(P|M) = \dfrac{200}{250} \cdot \dfrac{140}{200} = \dfrac{140}{250} = .56$

(g) $P(F \text{ and } O) = \dfrac{50}{250} \cdot \dfrac{25}{50} = \dfrac{25}{250} = .10$

(h) $P(P|M) = \dfrac{140}{200} = .70$

(i) $P(O|F) = \dfrac{25}{50} = .50$

5

Populations, Samples, and Distributions

Major Topics and Key Concepts

5.1 Populations

- Population or Universe - the group to which the conclusions are to be generalized.
- Infinite or finite.

5.2 Samples

- Random sample - every item in the population has equal opportunity of being selected.
- Sampling with and without replacement.

5.3 Random Variables

- Discrete - only specific values can occur.
- Continuous - any value within an interval can occur.

5.4 General Probability Distributions for Discrete Random Variates

- Sum of probabilities of each specific value must equal one.

5.5 General Probability Distributions for Continuous Random Variates

- Total area under the frequency curve is always one.
- Probability of any one specific value is zero.

5.6 Expected Values of Random Variables

- The mean, μ, of a probability distribution.
- The variance, σ^2, of a probability distribution.
- Standardizing a random variable.

5.7 Sets of Random Variables

5.8 Specific Discrete Probability Distributions - (Sections 5.8 through 5.11)

5.9 The Binomial Distribution

- For independent trials in which each trial has only two possible outcomes.
- Used if sampling is with replacement or from infinite population.
- X = a count of the number of occurrences in n trials.
- Use equation 5.13 or Table III.

5.10 The Hypergeometric Probability Distribution

- For dependent trials in which each trial has only two possible outcomes.
- Used if sampling without replacement.
- X = a count of the number of occurrences in n trials.
- Use equation 5.18 or 5.19.

5.11 The Poisson Probability Distribution

- X = a count of the number of occurrences per interval or unit.
- Use equation 5.20 or Table IV.

5.12 Normal Probability Distributions

- A continuous probability distribution.
- Two parameters: mean, μ, and variance, σ^2.
- Standardized Normal Distribution has $\mu = 0$, $\sigma^2 = 1$.
- $Z = \frac{X - \mu}{\sigma}$ transforms X, a normal random variable, which has mean μ and standard deviation σ into a Standard Normal Distribution.
- Use of Table V for the Standard Normal Distribution.

5.13 Summary

CHAPTER 5: Review Test

1. In a broad sense, a set or collection is a statistical:
    (a) Element.
    (b) Universe.
    (c) Distribution.
    (d) Process.

2. The things of which the population is composed are called:
    (a) Its elements.
    (b) The universe.
    (c) Finite.
    (d) Infinite.

3. An experiment which involves the analysis of a manufacturing process will most likely involve:
    (a) A finite population.
    (b) A binomial distribution.
    (c) An infinite population.
    (d) Discrete variables.

4. Which of the following is a random event?
    (a) The winner of last year's Irish sweepstakes.
    (b) The winning time in the 100-meter dash in the 1976 Olympic Games.
    (c) The winner of next year's World Series.
    (d) The distance from Chicago O'Heare Airport to the Dallas-Fort Worth Airport.

5. Which of the following is not a random event?
    (a) The actual weight of a randomly selected sack of Idaho potatoes labeled "10 pounds" on the sack.
    (b) The flying time for the next flight from San Antonio, Texas to Tuscaloosa, Alabama.
    (c) The price of a gallon of gasoline at a randomly selected station in Fresno, California.
    (d) The amount of rainfall in the Catskill Mountains in 1978.

6. The investigation of a small sample of a whole population is often justified because:
    (a) It is more economical.
    (b) Elements may be destroyed in the testing process.
    (c) It is impossible to examine all elements.
    (d) All of the above.

7. A sample is random if:
   (a) A table of random numbers is used to select the sample elements.
   (b) It is selected in such a way that all elements have the same probability of being chosen.
   (c) The property of discreteness is inherently present in the sample.
   (d) Both (a) and (b).

8. Random sampling is sometimes called:
   (a) Continuous sampling.
   (b) Independent sampling.
   (c) Sampling without replacement.
   (d) Chance selection.

9. The term representing the outcome of an experiment before the experiment is performed is a:
   (a) Random variable.
   (b) Dependent variable.
   (c) Finite variable.
   (d) Hypothetical probability.

10. A random variable is _____ if there is a definite distance from any possible value of the random variable to the next possible value.
    (a) Continuous.
    (b) Discrete.
    (c) Dependent.
    (d) Finite.

11. A continuous random variable cannot be described by a list of probabilities, but rather by a continuous curve. This is because:
    (a) It is impossible to list the probabilities.
    (b) Such a variable takes on any given value with probability zero.
    (c) Both of the above.
    (d) None of the above.

12. The distribution of any continuous random variable is specified by a curve called a:
    (a) Frequency curve.
    (b) Density curve.
    (c) Geometric curve.
    (d) Probability curve.

13. For a continuous random variable, the probability that, when an experiment is carried out, X will assume a value in a given interval equals:

    (a) Zero.
    (b) The probability density.
    (c) 1.0— (the area under the curve and over that interval).
    (d) The area under the curve and over that interval.

14. The expected value for a discrete random variable X is:

    (a) The mean of X.
    (b) The weighted mean of the possible values of X.
    (c) What X is expected to be in an average sense.
    (d) All of the above.

15. If X has a mean μ and standard deviation σ, subtracting μ from X and dividing by σ:

    (a) Gives the mean a value of one.
    (b) Standardizes the variable.
    (c) Makes the standard deviation equal to two.
    (d) All of the above.

16. Var (X + Y) = Var (X) + Var (Y) if:

    (a) X and Y are dependent.
    (b) X and Y are independent.
    (c) X and Y are not random variables.
    (d) X and Y are both positive.

17. A binomial population is one in which the elements may be split into two classes that are conventionally labeled _____ and _____.

    (a) Defective, nondefective.
    (b) Success, failure.
    (c) p, (1-p).
    (d) p, np.

18. The binomial distribution is symmetric if:

    (a) p = 0.
    (b) p = 1.
    (c) p = .25.
    (d) p = .5.

19. The binomial distribution could be used to calculate the probability of exactly r successes in n trials when

    (a) The trials are all independent of each other.
    (b) The probability of a success changes for each trial.
    (c) The number of trials is random.
    (d) The random variable is continuous.

20. The hypergeometric distribution could be used to calculate the probability of exactly r successes in n trials when:

    (a) The sampling is done without replacement.
    (b) The numbers of possible successes and failures in the population are known.
    (c) The population is finite.
    (d) All of the above.

21. The Poisson distribution could be used to calculate the probability of exactly r occurrences per unit when:

    (a) The population is finite.
    (b) The sample is gathered without replacement
    (c) The expected number of occurrences per unit is specified.
    (d) All of the above.

22. A normal frequency curve is:

    (a) A continuous probability distribution.
    (b) Symmetrical.
    (c) Bell-shaped.
    (d) All of the above.

23. Normal distribution tables are based on the standard normal distribution which is defined as that normal distribution which has:

    (a) Mean 0, Variance 0.
    (b) Mean 0, Any Variance.
    (c) Mean 0, Variance 1.
    (d) Variance 0, Any Mean.

24. If we desire to know the area under the normal curve between two values a and b (a and b on the same side of the origin), we _____ the corresponding areas in the normal table.

    (a) Subtract.
    (b) Add.
    (c) Divide.
    (d) Multiply.

25. If the average number of phone calls arriving at a switchboard is 5 per 15-minute interval, then to calculate the probability of exactly r calls during a randomly selected 15-minute interval, use the

    (a) Binomial distribution.
    (b) Hypergeometric distribution.
    (c) Poisson distribution.
    (d) Normal distribution.

26. From a group known to consist of 15 males and 10 females, a random sample of 5 will be selected without replacement. To calculate the probability of exactly r males being selected, use the:

(a) Binomial distribution.
(b) Hypergeometric distribution.
(c) Poisson distribution.
(d) Normal distribution.

CHAPTER 5: Answers to Review Test

Question	Answer	Text Section Reference
1	b	5.1
2	a	5.1
3	c	5.1
4	c	5.3
5	d	5.3
6	d	5.2
7	d	5.2
8	c	5.2
9	a	5.3
10	b	5.4
11	c	5.5
12	a	5.5
13	d	5.5
14	d	5.6
15	b	5.6
16	b	5.7
17	b	5.9
18	d	5.9 and Table III
19	a	5.9
20	d	5.10
21	c	5.11
22	d	5.12
23	c	5.12
24	a	5.12
25	c	5.11
26	b	5.10

CHAPTER 5: Review Problems

1. Given that the random variable X has the following probability distribution:

r	0	1	2	3	4
P(X = r)	.50	.20	.15	.10	.05

   (a) Find the probability that X is even.
   (b) Find the mean and variance of X.

2. If X follows a binomial distribution with n = 5, p = .2, then
   (a) Find $P(X \leq 3)$ using the table.
   (b) Find $P(X = 3)$ using the formula and table.
   (c) Find the probability that X is not as large as 3.
   (d) Find the probability that X is at least 2.

3. If X follows a Poisson distribution with θ = 2.6, then
   (a) Find $P(X = 3)$ using the table.
   (b) Find $P(X \leq 2)$ using the table.
   (c) Find $P(X > 1)$ using the table.

4. A company has twenty-five employees. Twenty of the employees say they like to attend the summer musical series, and five do not. The company is planning to give away three season tickets to the musicals.

   (a) What is the probability that all three tickets will be given to those who want them if the tickets are given away in a random drawing?
   (b) Find the probability that exactly one season ticket will go to someone who does not enjoy the musicals.

5. The surface of an automobile is approximately 200 square feet. Historically, records at the Stutz Bearcat Motor Company show that there are about 0.2 paint blemishes per square foot of painted surface. Find the probability that a new car will have less than three paint blemishes.

6. A survey conducted by a national ski magazine shows that 30% of their subscribers have skied in Utah. The Utah Travel Council would like to interview skiers and determine their attitudes toward Utah skiing. An employee of the Travel Council has randomly selected 10 names from the ski magazine's subscription list.

   (a) Find the probability that none of the 10 have skied in Utah.
   (b) Find the probability that no more than 3 have skied in Utah.

(c) Find the probability that two or fewer have not skied in Utah.
(d) Find the probability that exactly four have skied in Utah.

7. If Z follows a standard normal distribution, then find
   (a) $P(Z < 1.5)$.
   (b) $P(-1.2 < Z < 1.6)$.
   (c) $P(Z < -1.05)$.
   (d) C such that $P(Z < C) = .1093$.

8. The produce trucks which arrive at the Ptomaine Tavern distribution centers carry a mean weight of 3.2 tons of produce. The standard deviation of the weights for individual truck loads is 800 pounds. What is the probability that, if the weights follow a normal distribution, the next arriving truck will have
   (a) 7,600 pounds or more of produce in it?
   (b) Between 7,000 pounds and 7,640 pounds of produce?
   (c) Exactly 6,000 pounds of produce?
   (d) Less than 6,400 pounds?
   (e) Find the weight such that 9% of all trucks carry less weight in produce than this amount.

9. A numeric control (computer operated) milling machine at the Jet Dynamics Laboratories is used to make bolts which hold engines on the wings of wide bodied aircraft. When the milling machine is working properly, these come off the machine 33 millimeters thick. These diameters are normally distributed with a standard deviation of .20 millimeters. The thickness of these bolts are checked very carefully since the strength of the bolt is directly related to the diameter.
   (a) Find the probability that an individual, randomly selected bolt will be more than 32.5 millimeters in diameter.
   (b) What is the probability that a bolt will be between 32.8 and 33.40 millimeters in diameter?
   (c) Find the symmetric region such that 88% of all engine support bolts will have diameters within these bounds.

51

CHAPTER 5: Solutions to Review Problems

1. <u>Solution</u>

   a. $P(X \text{ is even}) = P(X=0) + P(X=2) + P(X=4) = .70$

   b. $\mu = \Sigma r \cdot P(X=r) = 0 \cdot (.5) + 1 \cdot (.20) + 2 \cdot (.15) + 3 \cdot (.10)$
   $+ 4 \cdot (.05) = 1.0$

   $\sigma^2 = \Sigma(r-\mu)^2 P(X=r)$
   $= (0-1)^2 \cdot (.50) + (1-1)^2 \cdot (.20) + (2-1)^2 \cdot (.15)$
   $+ (3-1)^2 \cdot (.10) + (4-1)^2 \cdot (.05) = 1.5$.

   NOTE: These formulae are for <u>discrete</u> random variables only.

2. <u>Solution</u>

   As a guide to using the binomial table, list the sample space and underline the points that you are interested in. Remember that the binomial table is CUMULATIVE.

   a. Enter Table III with
   $n = 5$, $p = .20$, $r = 3$ $r|0,1,2,3,4,5$
   $P(X \leq 3) = .9933$

   b. $P(X = 3) = P(X \leq 3) - P(X \leq 2)$ $r|0,1,2,\underline{3},4,5$
   $= .9933 - .9421 = .0512$

   or $P(X = r) = \binom{n}{r} \cdot p^r(1-p)^{n-r} = \binom{5}{3} \cdot (.2)^3(.8)^2$
   $= 10 \cdot (.008) \cdot (.64)$
   $= .0512$

   c. $P(X < 3) = P(X \leq 2) = .9421$ $r|0,1,2,3,4,5$

   d. $P(X \geq 2) = 1 - P(X \leq 1)$ $r|0,1,2,3,4,5$
   $= 1 - .7373 = .2627$

3. <u>Solution</u>

   a. Enter Table IV with $\theta = 2.6$, $r = 3$
   $P(X = 3) = .2176$

   b. $P(X \leq 2) = P(X = 0) + P(X = 1) + P(X = 2) = .5184$

c. $P(X > 1) = 1 - P(X = 0) - P(X = 1) = .7326$

Remember, the Poisson table is NOT CUMULATIVE like the binomial table.

4. **Solution**

Since one employee can be given at most one season ticket, sampling is <u>without</u> replacement. A <u>success</u> is defined as a person who would like a <u>ticket</u> receiving one. The <u>hypergeometric</u> distribution must be used: $a = 20$, $b = 5$, $n = 3$.

a. $P(X = 3) = \dfrac{\binom{a}{3} \cdot \binom{b}{n-3}}{\binom{a+b}{n}} = \dfrac{\binom{20}{3} \cdot \binom{5}{0}}{\binom{25}{3}} = \dfrac{1140 \cdot 1}{2300} = .4957$

b. If exactly one of three does not enjoy the musicals, then exactly two of three must enjoy them:

$P(X = 2) = \dfrac{\binom{20}{2}\binom{5}{1}}{\binom{25}{3}} = \dfrac{190 \cdot 5}{2300} = .4130$.

5. **Solution**

Since we have a <u>rate</u> of occurrences, X follows a Poisson distribution:

$P(X < 3) = P(X = 0) + P(X = 1) + P(X = 2) = .9988$.

6. **Solution**

Since $n = 10$ is fixed, $p = .30$ for all subscribers, and X = count of the number of subscribers <u>who have skied</u> in Utah, then X follows a binomial distribution.

a. $P(X = 0) = .0282$ from Table III; $n = 10$, $p = .30$, $r = 0$.

b. P(no more than 3) = $P(X \leq 3)$ $r | 0, 1, 2, 3, \ldots, 10$
= .6496

c. P(two or fewer have <u>not</u>) $r | 0, 1, 2, \ldots, 7, \underline{8, 9, 10}$
= Pr(8 or more have)
= $1 - P(X \leq 7) = 1 - .9984$
= .0016

d. P(exactly four) = P(X ≤ 4)   r ..,3,<u>4</u>,5,...,10
    − P(X ≤ 3) = .8497 − .6496
    = .2001 .

7. <u>Solution</u>

There are basically two type problems associated with the standard normal tables: 1, given an interval on the Z axis, find the area under the curve, or 2, given an area, define the interval on the Z axis. Always draw a picture of the normal probability curve; mark the interval of interest on the Z axis and shade in the desired or given area under the curve. Develop a habit of doing this.

a. P(Z < 1.5) = .5 + .4332
    = .9332

The table gives only the area between the mean and indicated value on the axis.

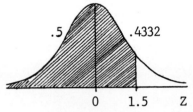

b. P(−1.2 < Z < 1.6) = .3849
    + .4452 = .8301

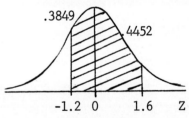

c. P(Z < −1.05) = .5 − .3531
    = .1469

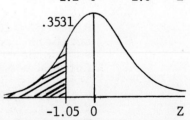

d. This is the second type of problem; given a probability, you must determine the interval that satisfies the conditions. The area to the left of C must be .1093; therefore, C must be negative. Subtract .1093 from .5 to obtain an area compatible with the table. Enter Table V with an area of

$P(z<c) = /0.93$

.3907 and read the Z value in the margins. The required value is C = -1.23; hence,
P(Z < -1.23) = .1093.

8. **Solution**

Using the formula, $Z = \frac{X-\mu}{\sigma}$, any normal variable can be transformed into a standard normal variable. This must be done before using Table V.

a. P(X > 7600) = P$(Z > \frac{7600-6400}{800})$

= P(Z > 1.5)

= .5 - .4332 = .0668

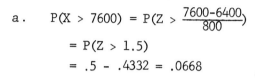

b. P(7000 < X < 7640)

= P$(\frac{7000-6400}{800} < Z < \frac{7640-6400}{800})$

= P(.75 < Z < 1.55)

= .4394 - .2734 = .1660

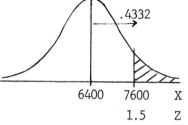

c. P(X = 6000) = 0

d. P(X < 6400) = P(Z < 0) = .5

e. First, solve for the value on the Z axis, transform back to obtain the boundary on the X axis. The value of C must be negative since the area to its left is less than .5. Enter Table V with an area of .41 and read 1.34; therefore, C = -1.34. To complete the solution, solve for X

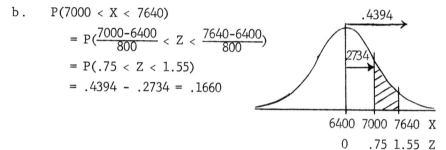

$-1.34 = \frac{X-6400}{800}$ or X = 5328

Hence, 9% of the trucks carry less than 5328 pounds of produce to the Ptomaine Tavern distribution centers.

55

9. <u>Solution</u>

a. $P(X > 32.5) = P(Z > \frac{32.5-33}{.20})$
    $= P(Z > -2.5)$
    $= .4938 + .5 = .9938$

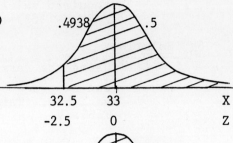

b. $P(32.8 < X < 33.40)$
    $= P(-1.0 < Z < 2.0)$
    $= .3413 + .4772 = .8185$

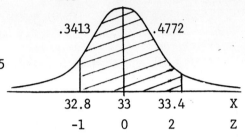

c. Solve on the Z axis first.
    $P(-C < Z < C) = .88$
    $C = 1.555$ from Table V.
    Therefore, to find bounds on the X-axis, transform back
    $\pm 1.555 = \frac{X-33}{.2}$
    $X = 33 \pm .311$.

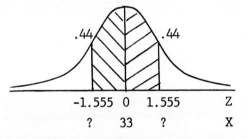

Hence, 88% of all bolts have diameters between 32.689 and 33.311 millimeters.

56

Probability Approximations

Major Topics and Key Concepts

- First, determine from the nature of the problem what the correct distribution of the random variable is.

- Use an approximation if it is appropriate and more convenient than calculating exact probabilities.

6.1 The Binomial Approximation of the Hypergeometric

- Appropriate if Hypergeometric is the correct distribution (see Section 5.10) and
- If the population size is large relative to the sample size (if $n < .10N$).
- Procedure: Let probability of success for the binomial be

$$p = \frac{b}{a+b} = \frac{\text{\# of successes in the population}}{\text{total \# of objects in the population}}$$

- Use binomial distribution (equation 5.13) or Table III (see Section 5.9).

6.2 The Poisson Approximation of the Binomial

- Appropriate if binomial is the correct distribution (see Section 5.9) and
- If the sample size, n, is large and the probability of success, p, is very small or very large.
- Procedure: let $\theta = n \cdot p$.
- Use Poisson distribution (equation 5.20) or Table IV.

6.3 The Normal Approximation to the Binomial

- Appropriate if binomial is the correct distribution (see Section 5.9) and
- If $np(1-p) \geq 5$.
- Procedure: let $\mu = np$ and $\sigma = \sqrt{np(1-p)}$. (Include corrections for continuity.)
- X is approximately normal ($\mu = np$, $\sigma^2 = np(1-p)$).
- Use Table V.

CHAPTER 6: Review Test

1. Which of the following is not a useful approximation?

    (a) Hypergeometric approximated by Binomial.
    (b) Binomial approximated by Poisson.
    (c) Hypergeometric approximated by Normal.
    (d) Normal approximated by Poisson.

2. The hypergeometric distribution can be approximated with the binomial:

    (a) When the number of successes in the population is nearly equal to the number of failures in the population.
    (b) When the sample is sufficiently large relative to the population size.
    (c) When the population size is large, since the probability of success would change very little when sampling without replacement.
    (d) Because the word "binomial" has fewer letters than "hypergeometric."

3. When using the binomial to approximate the hypergeometric, the probability of success that should be used is:

    (a) $p = a/b$.
    (b) $p = n/(a+b)$.
    (c) $p = a/(a+b)$.
    (d) $p = .5$.

4. The Poisson can be used to approximate the binomial when

    (a) The probability of success, P, is near .5 and n is large.
    (b) p is near 1.0 or near zero and n is large.
    (c) n is large and p is near zero.
    (d) $np \geq 5.0$.

5. When approximating the binomial distribution, the parameter of the Poisson distribution, θ, is estimated by using

    (a) p or q, whichever is smaller.
    (b) np or nq, whichever is smaller.
    (c) n.
    (d) $\sqrt{np(1-p)}$.

6. A rule of thumb for determining if the Poisson approximation for the binomial is good is if

    (a) Either $p \leq .10$ or $q \leq .10$.
    (b) $n \geq 120$.

(c) Either np ≤ 5.0 or nq ≤ 5.0.
(d) Both (a) and (c).

7. The normal distribution provides an adequate approximation for the binomial distribution when:

   (a) p is extreme and n is large.
   (b) p is near 0.5 regardless of the value of n.
   (c) np(1-p) ≥ 5.
   (d) Both (b) and (c) are correct.

8. Further accuracy can be achieved with normal approximations for the binomial when:

   (a) A continuity correction is used.
   (b) A rough sketch of the curve and area desired is made.
   (c) np = 100 and $np\sqrt{(1-p)}$ = 1.
   (d) Binomial and normal tables are averaged together.

9. In general, to apply the continuity correction, for a ≤ X ≤ b, we _____ from the lower value on the X scale and _____ to the upper value.

   (a) Add 1, subtract 1.
   (b) Add 1/2, subtract 1/2.
   (c) Subtract 1/2, add 1/2.
   (d) Subtract 1, add 1.

10. A firm estimates that 4 percent of its accounts receivable cannot be collected. What is the probability that out of its 150 current accounts receivable, eight or more will be uncollectible? (Use an approximation.)

    (a) .7340   (c) .2660
    (b) .1492   (d) .2340

11. The union in a plant claims that only 20 percent of the workers are opposed to a strike. If a random sample of 225 workers is taken and if the union claims are correct, what is the probability of obtaining between 54 and 64 opponents inclusive of the strike in the sample?

    (a) .5006   (c) .4994
    (b) .0772   (d) .4222

12. If only 0.5% of the integrated circuits produced by Omaha Instruments are defective, what is the probability of 4 or less defectives appearing in a shipment of 640 circuits?

    (a) .7669   (c) .1781
    (b) .7806   (d) .3975

CHAPTER 6: Answers to Review Test

Question	Answer	Text Section Reference
1	d	Summary Table
2	c	6.1
3	c	6.1
4	b	6.2
5	b	6.2
6	d	6.2
7	c	6.3
8	a	6.3
9	c	6.3
10	c	Use Normal with correction for continuity
11	b	Use Normal with correction for continuity
12	b	Use Poisson, $\theta = 3.2$

CHAPTER 6: Review Problems

1. Find the approximate probability that X = 6 if X follows the hypergeometric distribution,

$$P(X=r) = \frac{\binom{100}{r}\binom{400}{25-r}}{\binom{500}{25}} .$$

2. If X follows a binomial distribution with n = 200 and p = .004, find the approximate probability that X is less than or equal to 2.

3. If X follows a binomial distribution with n = 150 and p = .400, find the approximate probability that X is at least 50 but not more than 58.

4. (a) If three cards are dealt from a deck of 52, what is the probability of at least two ACES?

   (b) Would you expect this approximation to be very close?

5. If X follows a hypergeometric distribution and there are 30 favorable outcomes and 4970 unfavorable outcomes in the population, find the probability that X is at least one if the sample size is n = 100.

6. A major producer of steel belted radial tires has had a problem with "belt-edge separation at high mileage." The result was several accidents attributable to tire defects. An industry report in 1977 showed that in one year the return rate was 27%. If a sample of 500 customers of this type of tire was taken, what is the probability that at least 120 customers would have had to return their tires because of defects?

7. A CPA firm has 200 clients, 150 of which are located in-state. If 10 clients are randomly selected, find the probability that exactly eight will be located in-state.

8. A particular type of birth control device is successful 98.5% of the time when used properly. If a random sample of 240 women use this device for one year, find the probability that the device accomplishes its mission in every case.

CHAPTER 6: Solutions to Review Problems

1. <u>Solution</u>

   Evaluating the exact solution would be rather tedious. Since the sample size, n = 25, is less than 10% of the population size, a + b = 500, the binomial approximation can be used. In a population consisting of 500 members, 100 members have the trait that is being counted; hence, the probability of the first member selected at random having the trait is p = 100/500 = .20. Therefore, X is approximatly distributed as binomial (n = 25, p = .20).

   $$P(X = 6 | n = 25, p = .20) = P(X \leq 6) - P(X \leq 5)$$
   $$= .7800 - .6167 = .1633 \quad \text{from Table III.}$$

   (NOTE: Exact probability use hypergeometric is .1671).

2. <u>Solution</u>

   Because n is large, p is small and np < 5, use the Poisson distribution for approximation. Let θ = np = .8. Hence,

   $$P(X \leq 2 | \theta = .8) = P(X = 0) + P(X = 1) + P(X = 2)$$
   $$= .4493 + .3595 + .1438 \quad \text{from Table IV}$$
   $$= .9526 .$$

3. <u>Solution</u>

   Since np(1-p) = 36 > 5, we can use the normal approximation.

   $$\mu = np = 60 \text{ and } \sigma = \sqrt{np(1-p)} = 6.0.$$

   We are using a continuous distribution to approximate a discrete distribution; since we want to include the lower limit, 50, use 49.5; to include the upper limit, 58, use 58.5.

   $$P(50 \leq X \leq 58 | \text{binomial}) \doteq P(49.5 \leq X \leq 58.5 | \text{normal}).$$
   $$= P\left(\frac{49.5 - 60}{6} < Z < \frac{58.5 - 60}{6}\right)$$
   $$= P(-1.75 < Z < -.25) = .4599 - .0987 = .3612.$$

4. **Solution**

   a. Since sampling is without replacement, X, the count of number of ACES, follows a hypergeometric distribution. Because the sample size, n = 3, is less than 10% of the population size, a + b = 52, the binomial approximation can be used; n = 3, p = 4/52.

   $$P(X \geq 2 | \text{hypergeometric}) \doteq P(X \geq 2 | \text{binomial})$$
   $$= P(X = 2) + P(X = 3)$$

   Since p = 1/13 is not tabulated, use formula 5.13.

   $$P(X \geq 2) = \binom{3}{2}(\tfrac{1}{13})^2 (\tfrac{12}{13})^1 + \binom{3}{3}(\tfrac{1}{13})^3 (\tfrac{12}{13})^0$$
   $$= .0164 + .0005 = .0169 .$$

   b. The population size is not large even though the 10% rule is met. The true probability is .01321.

5. **Solution**

   Because the sample size, n = 100, is less than 10% of the population size, a + b = 5000, the binomial approximation can be used with n = 100 and p = 30/5000. But since p = .006 is very small and np = 0.6 is less than 5, the Poisson approximation can be used for the binomial, with θ = np = 0.6. Therefore,

   $$P(X \geq 1 | \text{hypergeometric}) \doteq P(X \geq 1 | \text{Poisson})$$
   $$= 1 - P(X = 0) \quad \quad (\text{Table IV})$$
   $$= .4512.$$

6. **Solution**

   The random variable, X = a count of the customers that returned tires, follows a binomial distribution, n = 500 and p = .27. Since np(1-p) = 98.55 is greater than 5, a normal approximation is appropriate, μ = 500(.27) = 135, σ = √98.55 = 9.927. The lower bound, 120, is included in the region of interest; therefore use 119.5 in order to correct for continuity.

   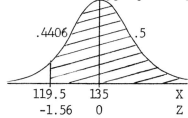

   .4406   .5

   119.5   135   X
   -1.56   0   Z

$$P(X \geq 120 | \text{binomial}) \doteq P(X \geq 119.5 | \text{normal})$$

$$= P(Z \geq \frac{119.5 - 135}{9.927})$$

$$= P(Z \geq -1.56) = .4406 + .5 = .9406.$$

7. Solution

If sampling is without replacement (meaning that the same client could not be selected more than once), the hypergeometric is the correct distribution to use. But $n = 10 < .10 \cdot 200$, the binomial approximation can be used with $n = 10$, $p = 150/200 = .75$.

$$P(X = 8 | \text{hypergeometric}) = P(X = 8 | \text{binomial})$$

$$= \binom{10}{8} \cdot .75^8 (.25)^2 = 45(.1001129) \cdot (.0625)$$

$$= .2816 \; .$$

8. Solution

The population size is infinite; X is a count of the number of times the device is effective.

We want $P(X = 240 \; n = 240, P = .985)$. Here $np = 236.4$ and $n(1-p) = 3.6$. Since the smaller of the two is less than 5.0, the Poisson approximation is appropriate; however, we must count the number of failures instead of successes, i.e., 240 successes means 0 failures. Now, $\theta = n(1-p) = 3.6$.

$$P(X = 0 | \text{failures} | \theta = 3.6) = .0273 \quad \text{(from Table IV)}$$

Sampling Distributions

Major Topics and Key Concepts

The sample mean, $\bar{X}$, is a statistic;
A statistic is a random variable;
Random variables have probability distributions;
Therefore, $\bar{X}$ must have a probability distribution.
What is it?

7.1 Sampling and Inference

- A <u>parameter</u> is a number which describes a characteristic of a population.
- A <u>statistic</u> is a random variable whose value is derived from a sample.
- <u>Inference</u> is generalizing the results of a sample to entire populations; that is, using a statistic to estimate a parameter.

7.2 Expected Values and Variances

- $E(\bar{X}) = \mu \implies$ the probability distribution of $\bar{X}$ will always have the same mean value as the population from which the sample was taken.
- Variance $(\bar{X}) = \sigma^2/n \implies$ the probability distribution of $\bar{X}$ will always have a smaller variance than that of the (infinite) population from which the sample was taken--smaller by a factor of $1/n$.
- Standard deviation $(\bar{X}) = \sigma/\sqrt{n}$.
- Conclusion: The population of all possible $\bar{X}$ values will have a higher frequency of occurrence near the true population mean, μ, than the original population will have. This will happen regardless of the distribution of the original population.

7.3 Sampling from Normal Populations

- If the original population from which a sample is taken is <u>normal</u> with mean, μ, and variance, σ^2, then the <u>distribution</u> of all possible values of $\bar{X}$ is normal with mean, μ, and variance, σ^2/n.
- This is true for any sample size n.

7.4 The Standardized Sample Mean

- If the original population is normal (μ, σ^2), then $Z = \dfrac{\bar{X} - \mu}{\sigma/\sqrt{n}}$ is standard normal.
- Table V can be used to determine probabilities for $\bar{X}$.

7.5 The Central Limit Theorem

- Whatever the distribution of the original population, if n is large, then the distribution of all possible values of $\bar{X}$ is approximately normal with mean, μ, and variance σ^2/n.
- Also, $Z = \dfrac{\bar{X} - \mu}{\sigma/\sqrt{n}}$ is approximately standard normal.
- In terms of the sample total,

  $Z = \dfrac{\Sigma X - n\mu}{\sigma \cdot \sqrt{n}}$ is approximately standard normal.
- "Large" generally implies n > 30.

7.6 Sample Results from Two Populations

- $d = \bar{X}_1 - \bar{X}_2$ denotes the distance between two sample means.
- $\mu_d = E(\bar{X}_1 - \bar{X}_2) = \mu_1 - \mu_2$.
- $\sigma_d = \sqrt{\dfrac{\sigma_1^2}{n_1} + \dfrac{\sigma_2^2}{n_2}}$.
- If both populations are normal, then the distribution of $(\bar{X}_1 - \bar{X}_2)$ is normal.
- If the two original populations are not normal, then the distribution of $(\bar{X}_1 - \bar{X}_2)$ is approximately normal if both sample sizes are large.

7.7 Two Samples from Binomial Populations

- A generalization of Section 6.3.
- X_1 = number of successes in n_1 trials, $f_1 = X_1/n_1$.
- X_2 = number of successes in n_2 trials, $f_2 = X_2/n_2$.
- $(f_1 - f_2)$ is the distance between two sample proportions.

- $(f_1 - f_2)$ is approximately normal with mean $(p_1 - p_2)$ and variance $p_1(1 - p_1)/n_1 + p_2(1 - p_2)/n_2$.
- See equation 7.15.

7.8 Summary

- Review Table 7.3 carefully!

CHAPTER 7: Review Test

1. A sample may be drawn to:
    (a) Save needless waste of time, money, and effort.
    (b) Discover facts about a population.
    (c) Make inferences about a parameter.
    (d) All of the above.

2. In general, a _____ is a number describing some aspect of a population.
    (a) Sample.
    (b) Parameter.
    (c) Inference.
    (d) Correction factor.

3. In making inferences about a parameter on the basis of a sample, we need knowledge of:
    (a) Statistics.
    (b) Computers.
    (c) Sampling distributions.
    (d) All of the above.
    (e) (a) and (c) only.

4. It is not possible to determine exact parameters for:
    (a) Any distribution.
    (b) A discrete, finite population.
    (c) A continuous population.
    (d) All of the above.

5. The expected value of the sample mean $\bar{X}$ _____ coincides with the population mean μ.
    (a) Always
    (b) Often
    (c) Seldom
    (d) Never

6. If the population is infinite, so that the elements in a random sample are independent, then the standard deviation of the population of all possible $\bar{X}$ values is:
    (a) $\dfrac{\sigma}{\sqrt{n}}$.
    (b) $\sigma \cdot \sqrt{n}$.
    (c) σ^2/n .
    (d) $\sigma^2/\sqrt{n}$.

7. If the population is normal, then $\bar{X}$ is:
   - (a) Approximately normally distributed with mean μ and variance σ^2/n.
   - (b) Exactly the same as the population mean.
   - (c) Represented by a slightly skewed histogram.
   - (d) Exactly normally distributed with mean μ and variance σ^2/n.

8. If the population has <u>any</u> distribution other than normal, then $\bar{X}$ has:
   - (a) The same distribution as the original population for large sample sizes.
   - (b) An unknown distribution regardless of the sample size.
   - (c) Has approximately a normal distribution for large sample sizes.
   - (d) Has exactly a normal distribution for any sample size.

9. The Central Limit Theorem states that if the sample size is large then:
   - (a) The standardized variable has approximately a normal distribution.
   - (b) $\bar{X}$ has approximately a normal distribution.
   - (c) Both (a) and (b).
   - (d) Neither (a) nor (b).

10. When comparing the sample means obtained from two independent samples, the statistic $d = \bar{X}_1 - \bar{X}_2$ will have a variance of:
    - (a) $\sigma_1^2 + \sigma_2^2$.
    - (b) $\sigma_1^2 - \sigma_2^2$.
    - (c) $\sigma_1^2/n_1 + \sigma_2^2/n_2$.
    - (d) $\sigma_1^2/n_1 - \sigma_2^2/n_2$.

11. The statistic $d = \bar{X}_1 - \bar{X}_2$ will have exactly a normal distribution with mean $\mu_1 - \mu_2$ if:
    - (a) Both original populations are normally distributed.
    - (b) Both sample sizes are greater than 30.
    - (c) Both (a) and (b).
    - (d) Neither (a) nor (b).

12. If neither of two populations is normally distributed and if one sample size is small, then:
    - (a) Pretend the statistic d is normal and continue with the solution.
    - (b) Punt.
    - (c) Obtain more data if possible.
    - (d) Let the computer solve it.

13. The normal approximation for the difference of two sample proportions can be made when both sample sizes are large because of the Central Limit Theorem.

   (a) True.
   (b) False.

CHAPTER 7: Answers to Review Test

Question	Answer	Text Section Reference
1	d	7.1
2	b	7.1
3	e	7.1
4	c	7.1
5	a	7.2
6	a	7.2
7	d	7.3
8	c	7.5
9	c	7.5
10	c	7.6
11	a	7.6
12	c	7.6
13	a	7.7

CHAPTER 7: Review Problems

1. If a population is normally distributed with mean of 16 and variance of 12, and if all possible samples of size n = 48 were drawn, then what is the distribution of all the sample means? Do you need the Central Limit Theorem to justify your answer?

2. In Mountain Dell Bank the average savings account balance is $285.50 with a standard deviation of $36.00. What is the probability that an audit of 324 accounts taken at random shows an average balance of $280.00 or less?

3. The distribution of weekly wages of illegal alien field workers is assumed to be normal with an average weekly wage of $80.00 and a standard deviation of $36.00.

   (a) If a random sample of 400 of these workers is taken, how many of them would you expect to have wages greater than $84.00 per week?
   (b) What is the probability that the mean weekly wage of the 400 sampled workers is greater than $84.00?

4. In the past, the mean claim size for a given auto insurance policy has been $1,354 and the standard deviation of the claim sizes has been $624. There is a backlog of 576 claims to be processed. What is the probability that the mean claim size $\bar{X}$ in the backlog exceeds $1,400?

5. A nationwide cosmetic company has some sales representatives working on commissions, others on salary. The mean monthly wages earned by the two groups are μ_1 = $350 and μ_2 = $320, respectively. The standard deviations are σ_1 = $120 and σ_2 = $32. If a random sample of 100 commissioned representatives and 64 salaried salespersons are taken:

   (a) What is the probability that the mean earnings of commissioned people will be less than the mean earnings of salaried personnel?
   (b) What is the probability that the average commissioned earnings will exceed the average salaried earnings by $20?

6. To determine the effectiveness of an advertising campaign, a random sample of 100 housewives will be asked if they had purchased baking soda in the last 3 months. After extensive advertising for 6 months, another

$d = X_1 - X_2$
$d = 125 - 125$
$d = 0$

random sample of 200 housewives will be asked the same question. Suppose that prior to the ads, $p_1 = 14\%$ of the population had purchased the product and after the ads $p_2 = 30\%$ had purchased baking soda in the 3-month period.
(a) State the distribution of the difference of the two sample proportions, $f_2 - f_1$.
(b) Find the probability that the proportion will be increased more than .20.

7. Suppose the distribution of systolic blood pressure for corporation vice presidents is normally distributed with mean of 125 and variance of 25, and for stock brokers it is also normal with a mean of 125 but a variance of 56.
(a) If one vice president and one stock broker are randomly selected, what is the probability that the systolic blood pressure of the stock broker will be more than 4 units higher than that of the vice president?
(b) If a random sample of size 20 is taken from each population, what is the probability that the mean for stock brokers will exceed that of vice presidents by more than 4 units?

$\bar{X} = 1354$
$\sigma = 624$
$n = 576$
$P(r > 1400)$

$\dfrac{1400 - 1354}{624} = \dfrac{46}{624} = .0731$

$.5 - .185 = .315$

$\sigma_d = \sqrt{\dfrac{\sigma_1^2}{n_1} + \dfrac{\sigma_2^2}{n_2}}$

$d_u = \bar{X}_1 - \bar{X}_2 = 350 - 320 = 30$

$\sqrt{\dfrac{120^2}{100} + \dfrac{32^2}{64}}$

$\sigma_d = \sqrt{144 + 16} = 12.64911$

$P(\bar{X}_s > \bar{X}_c) = P(d < 0)\, P\left(z < \dfrac{0 - 30}{12.64911}\right)$

$= -2.371$

$P\left(z < \dfrac{20 - 30}{12.64911}\right) = -10$

$= .7905$

-2371

$= .2852$

73

CHAPTER 7: Solutions to Review Problems

1. **Solution**

   Since X_i are Normal ($\mu = 16$, $\sigma^2 = 12$), then $\bar{X}$ is Normal ($\mu_{\bar{X}} = 16$, $\sigma^2_{\bar{X}} = 12/48$). The central limit is not needed if there is knowledge that the original population from which the sample is taken is normally distributed.

2. **Solution**

   Since n = 324 is large, the Central Limit Theorem allows us to claim that $\bar{X}$ is approximately normal, $\mu_{\bar{X}} = 285.50$,

   and $\sigma_{\bar{X}} = 36/\sqrt{324} = 2.0$.

   Hence, $Z = \dfrac{280 - 285.50}{2}$

   $= -2.75$.

   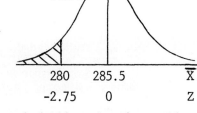

   From the normal tables, for Z = -2.75, read a probability of .4970; therefore, the desired probability in the tail is .5 - .4970 = .003.

3. **Solution**

   a. X is normal, $\mu = 80$, $\sigma = 36$. First, find the probability that one field worker earns more than $84.

   $Z = \dfrac{84 - 80}{36} = .11$.

   $P(X > 84) = P(Z > .11)$

   $= .5 - .0438$

   $= .4562$.

   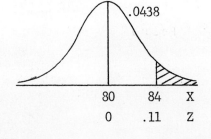

   Thus 45.62% of the population of workers have weekly wages over $84.00. If the sample is random and, therefore, representative of the population, then 45.62% of the sampled workers are in the same boat. This means:

   $$(.4562)(400) = 182.48$$

   or about 183 workers have wages over $84.00 per week.

b. The second part of this problem refers to sample means for samples of size n = 400. Since the wages in the population are normally distributed, samples taken from the population have normally distributed means (Section 7.3 of text). Even if the population wages were not normal, n = 400 is large and sample means for large samples are approximately normally distributed (the Central Limit Theorem).

The distribution of sample means is much narrower than the distribution of individual wages in part (a) above.

Its mean and standard deviation are:

$$\mu_{\bar{X}} = \mu = 80, \quad \sigma_{\bar{X}} = \frac{\sigma}{\sqrt{n}} = \frac{36}{\sqrt{400}} = 1.80 .$$

$$P(\bar{X} > 84) = P(Z > \frac{84 - 80}{1.80}) = P(Z > 2.22)$$

$$= .5000 - .4868 = .0132 .$$

Hence, it is unlikely that the mean weekly wage will be greater than $84.00.

4. **Solution**

We do not know if the claim size follows a normal distribution; but since n = 576 is large, we can claim $\bar{X}$ as approximately normal.

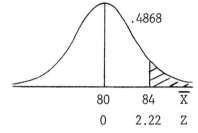

$\mu_{\bar{X}} = \$1354$,

$\sigma_{\bar{X}} = 624/\sqrt{576} = 26.$

$P(\bar{X} > 1400)$

$= P(Z > \frac{1400 - 1354}{26})$

$= P(Z > 1.77) = .5 - .4616 = .0384.$

5. <u>Solution</u>

We are not told the distributions of the two groups; however, since $n_1 = 100$ and $n_2 = 64$ are both large, $d = \bar{X}_1 - \bar{X}_2$ can be treated as a normal variable (Section 7.6).

$$\mu_d = \mu_1 - \mu_2 = \$30$$

$$\sigma_d = \sqrt{\frac{\sigma_1^2}{n_1} + \frac{\sigma_2^2}{n_2}}$$

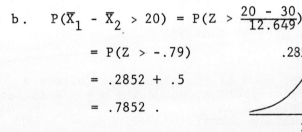

$$= \sqrt{\frac{120^2}{100} + \frac{32^2}{64}}$$

$$= \sqrt{160} = 12.649 \ .$$

a. $P(\bar{X}_1 < \bar{X}_2) = P(d < 0) = P(Z < \frac{0 - 30}{12.649})$

$\qquad = P(Z < -2.37)$

$\qquad = .5 - .4911$

$\qquad = .0089 \ .$

b. $P(\bar{X}_1 - \bar{X}_2 > 20) = P(Z > \frac{20 - 30}{12.649})$

$\qquad = P(Z > -.79)$

$\qquad = .2852 + .5$

$\qquad = .7852 \ .$

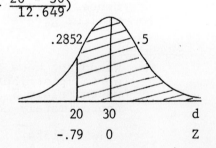

6. <u>Solution</u>

Here, $\mu_{f_2 - f_1} = p_2 - p_1 = .16$.

$$\sigma_{f_2 - f_1} = \sqrt{\frac{p_1(1 - p_2)}{n_1} + \frac{p_2(1 - p_2)}{n_2}} = \sqrt{.002254}$$

$$= .0475 \ .$$

The distribution of $f_2 - f_1$ is approximately normal; using equation 7.15,

$p(f_2 - f_1 < .20)$

$= p(Z > \frac{.20 - .16}{.0475})$

$= p(Z > .84)$

$= .5 - .2995$

$= .2005$.

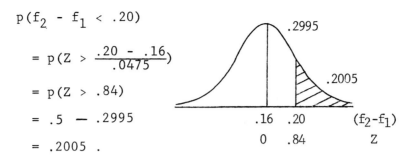

7. **Solution**

For vice presidents, $\mu_1 = 125$, $\sigma_1^2 = 25$. For stock brokers, $\mu_2 = 125$, $\sigma_2^2 = 56$. Since both populations are normally distributed, $d = \bar{X}_1 - \bar{X}_2$ is normal for any sample size.

a. Here, $n_1 = n_2 = 1$,

$\mu_d = \mu_1 - \mu_2 = 0$,

$\sigma_d = \sqrt{\frac{\sigma_1^2}{n_1} + \frac{\sigma_2^2}{n_2}} = 9$.

$P(\bar{X}_1 - \bar{X}_2 < 4)$

$= P(Z < \frac{4 - 0}{9})$

$= P(Z < .44) = .5 + .1700 = .6700$.

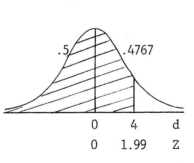

b. Here, $n_1 = n_2 = 20$,

$\mu_d = \mu_1 - \mu_2 = 0$,

$\sigma_d = \sqrt{\frac{25}{20} + \frac{56}{20}} = \sqrt{4.05}$.

$P(\bar{X}_1 - \bar{X}_2 < 4)$

$= P(Z < \frac{4 - 0}{2.01})$

$= P(Z < 1.99) = .5 + .4767 = .9767$.

Estimation

Major Topics and Key Concepts

- Inference - generalizing from a sample to a population or from a statistic to a parameter.

- Two types of inference

    Estimation - "I want some idea of what the parameter value is."
    Hypothesis testing - "I think I know what the parameter value ought to be; am I right?"

8.1 Estimators

- Two types: point estimators and confidence intervals.
- Properties of point estimators: unbiased, minimum variance.

8.2 Confidence Intervals for Normal Means: Known Variance

- $100(1-\alpha)\%$ CI for μ: $\bar{X} \pm Z_{\alpha/2} \cdot \sigma/\sqrt{n}$
- If n is small, the original population must have a normal distribution; otherwise, there are no restrictions on the original population.
- Meaning of a 95% confidence interval: if many different samples were taken from the same population, and if a 95% confidence interval were calculated for each sample, then we would expect that about 95% of all these different intervals would contain the true mean.

8.3 The t-Distribution

- Symmetric, zero mean, approaches a normal distribution as n approaches infinity.
- Use Table VI.
- Needed when true population variance, σ^2, is unknown.

8.4 Confidence Intervals for Normal Means: Unknown Variance

- $100(1-\alpha)\%$ CI for μ: $\bar{X} \pm t_{(\alpha/2, n-1)} \cdot s/\sqrt{n}$

- The original population must have a normal distribution.

8.5 Estimating Binomial p

- Point estimate of p: $f = \bar{X}/n$
- Exact 95% CI: Use Table VII.
- Approximate $100(1-\alpha)\%$ CI for p: $f \pm Z_{\alpha/2} \cdot \sqrt{f(1-f)/n}$

8.6 Sample Size

- Determining how large a sample is needed in order to insure that the confidence interval width will be no wider than some value.

8.7 Confidence Intervals for $\mu_1 - \mu_2$.

- Three necessary assumptions.
- Pooled estimate of common variance, equations 8.28, 8.29.
- Standard deviation of d where $d = \bar{X}_1 - \bar{X}_2$:
  equation 7.12 if both population variances are known.
  equation 8.31 if true variances are unknown but equal.
- $100(1-\alpha)\%$ CI for $(\mu_1 - \mu_2)$: equation 8.32

8.8 Two Samples From Binomial Populations

- $100(1-\alpha)\%$ CI for $(p_1 - p_2)$: equation 8.33

8.9 Summary

CHAPTER 8: Review Test

1. A sample quantity that serves to estimate an unknown parameter from a population is called:

    (a) An estimation.
    (b) An estimator.
    (c) An inference.
    (d) An hypothesis test.

2. An estimator must be _____ and depend _____ and not on the parameter to be estimated.

    (a) Unbiased; on the parameter
    (b) A specific value; on the method of measurement
    (c) A statistic; only on the sample
    (d) A parameter; only on the statistical sample

3. An estimator is said to be unbiased if:

    (a) It has an expected value equal to the parameter being estimated.
    (b) It always gives the parameter's value.
    (c) We divide by n-1 in calculating it.
    (d) On the average, it approximates the parameter's value.

4. As an estimator of σ (population standard deviation), a sample standard deviation s is:

    (a) Unbiased.
    (b) Extremely biased.
    (c) Somewhat biased.
    (d) None of the above.

5. A minimum variance unbiased estimator is an estimator which, in the first place, is unbiased and which, in the second place, has a _____ than any other unbiased estimator.

    (a) Smaller variance
    (b) Smaller standard deviation
    (c) Larger variance
    (d) Larger standard deviation

6. If the population is normal, $\overline{X}$ is the best estimator of μ because it is a minimum variance unbiased estimator.

    (a) True.
    (b) False.

7. In the equation $P(\bar{X} - .5 \leq \mu \leq \bar{X} + .5) = .90$, $\bar{X} - .5$ and $\bar{X} + .5$ are:

   (a) 90% confidence intervals for μ.
   (b) Random variables.
   (c) 90% confidence limits for μ.
   (d) Both (b) and (c).
   (e) Both (a) and (c).

8. If $\bar{X}$ in the equation of Question 7 is equal to 2.5, then we can say that there is a probability of .90 that the unknown μ lies between 2 and 3.

   (a) True.
   (b) False.

9. A 90% confidence interval on μ was found to be 21.0 to 24.4. Which of the following statements best interprets the meaning of this interval?

   (a) The probability that μ falls between 21.0 and 24.4 is .90.
   (b) If many samples were taken from the same population under similar conditions and if a confidence interval was constructed for each sample, then exactly 90% of the intervals would contain the true value of μ.
   (c) If many samples were taken from the same population under similar conditions and if a confidence interval was constructed for each sample, then I would expect 90% of the intervals to contain the true value of μ.
   (d) I expect that 90% of all possible value of μ fall between 21.0 and 24.4.

10. The probability that an 88% confidence interval contains the true parameter value is:

    (a) .88
    (b) Either zero or one since μ is non-random.
    (c) Indeterminable.
    (d) .5

11. The larger the number of degrees of freedom, the more closely the t distribution resembles a standard normal distribution.

    (a) True.
    (b) False.

12. For a fixed level of confidence and sample size less than 120, the value from the t-table will be _____ the value from the normal table.
    (a) Equal to
    (b) Less than
    (c) Greater than
    (d) Double

13. When using the normal approximation for constructing a confidence interval for a binomial p, n should be:
    (a) 50
    (b) 100
    (c) Large enough such that $np(1-p) \geq 5$.
    (d) Large enough such that $np \geq 5$.

14. The variance of f takes on its largest value when we assume that the population proportion is:
    (a) 1/4
    (b) 1/2
    (c) 1/8
    (d) 1/3

15. A 95% confidence interval as opposed to a 90% confidence interval:
    (a) Is narrower.
    (b) Gives greater precision.
    (c) Is wider.
    (d) Both (a) and (b).

16. If the sample size is increased, then confidence interval:
    (a) Is narrower.
    (b) Has more confidence.
    (c) Is wider.
    (d) Both (b) and (c).

17. If the population variance is not known, then the width of a confidence interval depends on the sample itself.
    (a) True.
    (b) False.

18. Some basic assumptions to be satisfied when using the techniques discussed in comparing samples from two normal populations include:
    (a) Independent random samples.
    (b) Common variance for both populations.
    (c) Equal sample sizes.
    (d) All of the above.
    (e) (a) and (b).

19. The pooled estimate of variance is:
    (a) The weighted mean of individual sample estimates of the variance.
    (b) The pooled sum of squares divided by the pooled degrees of freedom.
    (c) The best unbiased estimate of σ^2, the common variance when two population variances are equal.
    (d) All of the above.
    (e) (a) and (b) only.

20. The true variance of the difference between two sample means ($d = \bar{X}_1 - \bar{X}_2$) is:

    (a) $\dfrac{\sigma_1^2}{n_1} + \dfrac{\sigma_2^2}{n_2}$

    (b) $\dfrac{\sigma_1^2}{n_1} - \dfrac{\sigma_2^2}{n_2}$

    (c) $s^2(\dfrac{1}{n_1} + \dfrac{1}{n_2})$

    (d) $\sqrt{\dfrac{\sigma_1^2}{n_1} + \dfrac{\sigma_2^2}{n_2}}$

CHAPTER 8: Answers to Review Test

Question	Answer	Text Section Reference
1	b	8.1
2	c	8.1
3	a	8.1
4	c	8.1
5	a	8.1
6	a	8.1
7	d	8.2
8	b	8.2
9	c	8.2
10	b	8.2
11	a	8.3
12	c	8.4
13	c	8.5
14	b	8.5
15	c	8.6
16	a	8.6
17	a	8.6
18	e	8.7
19	d	8.7
20	a	8.7

CHAPTER 8: Review Problems

ONE SAMPLE PROBLEMS:

1. A soft drink machine is regulated so that the amount of drink dispensed is approximately normally distributed with a standard deviation equal to 0.6 ounces. What is the 95% confidence interval for the mean of all drinks dispensed by this machine if a random sample of 36 drinks had an average content of 7.5 ounces? Write your answer in the context of the problem.

2. A random sample of 9 quarts of Lowlife Ice Cream resulted in an average weight of 30.80 ounces per quart and a standard deviation of 0.63 ounces. Find a 98% confidence interval for the mean of all such quarts of ice cream, assuming the weights of quarts in the population follow the normal distribution.

3. Froz Handsoff is a strong contender for a gold medal in the ski jump event at the next Winter Olympics. In a sample of 5 jumps off the 90 meter hill Froz recorded leaps of 87, 85, 78, 82 and 84 meters. Construct a 90% confidence interval for the mean length of all his jumps from a 90 meter hill.

4. Due to recent arms limitation agreements the Army has been secretly testing the use of nuclear handgrenades. Due to their unorthodox shape and rather large kill radius, the Army wants to estimate the mean distance that a typical soldier could throw the grenade. A sample of 41 throws were measured in meters and the results were partially summarized:

$$\Sigma X_i = 881.5 \qquad \Sigma X_i^2 = 19873.85$$

Construct a 90% confidence interval.

5. In a random sample of new car purchasers in the midwest, the age of the buyers were recorded and tabulated. Construct a 95% confidence interval on the mean age.

Age of Buyer	Frequency
21 to 29	10
30 to 38	30
39 to 47	25
48 to 56	15
	80

6. The standard deviation of weekly wage rates for secretaries is known to be $27.90. A random sample of 100 secretaries yields a mean weekly wage rate of $157.45. What is the 92% confidence interval for the true mean weekly rate for the secretaries?

7. A large casualty insurance company is revising its rate schedules. A staff actuary wishes to estimate the average size of claim resulting from fire damage in apartment complexes having between ten and twenty units. He will use the current year's claim settlement experience as a sample. For buildings in this category there were 19 claim settlements. The average claim was $73,249, with a standard deviation of $37,246. What is the 90% confidence interval estimate of the mean claim size?

8. A sample poll of 100 citizens chosen at random revealed that 40 are "strongly against" using nuclear power as the primary source of energy in the future. Find a 95% confidence interval for the proportions of all citizens showing this position.

9. Another sample of 100 citizens chosen at random revealed that 82 believe it is impossible for the federal government to ever balance the budget. What are the 95% confidence limits for the proportion of all citizens that feel this way?

10. A major manufacturer of automobiles in the U.S. is genuinely concerned about feelings that car owners have toward the maintenance service provided by dealerships. The corporation conducted an extensive consumer confidence survey using a randomly selected nation-wide sample. Out of 1200 owners of American made cars, 420 responded that they were satisfied with the maintenance service provided by automobile dealerships. Using an 88% confidence interval, estimate the proportion of all owners of American made cars that are satisfied.

11. A department store wishes to estimate, with a confidence level of 98% and a maximum error of $5, the true mean dollar value of purchases per month by its customers with charge accounts. How large a sample should the store take from its records to satisfy the specifications if the standard deviation of monthly dollar purchases is known to be fifteen dollars?

12. A financial analyst wishes to estimate the mean value of total assets financed by long-term debt in a population consisting of many companies. He requires a 95% confidence interval with a width not to exceed $50,000. The specific value of the population standard deviation is unknown, but the analyst believes that it must be somewhere between $100,000 and $300,000. What sample size should the analyst use if he would rather have a sample which is too large than too small?

13. The manufacturer of a leading luxury car is interested in determining what proportion of drivers prefer the ride in his car to the ride in the "other leading luxury car." How many people should the manufacturer ask to rate the ride if it is desired that the sample proportion favoring its ride be accurate to within plus or minus three percent with 90% confidence? (Hint: Use the value of p for which Var(f) is maximum.)

14. A tire manufacturer found that in a random sample of 125 tires, eight tires were blemished. Find the 90% confidence limits for the proportion of blemished tires in the process.

15. ABC Company has just installed a new computer and is interested in determining just how long the computer takes to process the average job submitted to it. The output for 81 randomly selected jobs is examined and the mean is found to be 2.7 seconds and the standard deviation is 1.2 seconds. Find a confidence interval within which you are 90% sure the true mean processing time for jobs on the new computer will be.

16. The auditor of a bank found that in a sample of 100 demand deposit accounts 22 had errors of one kind or another. Find a 95% confidence interval for the proportion of checking accounts with errors. (Find the exact binomial probability interval.)

17. ABC's management is interested in finding out what proportion of its employees know how to program computers in order to determine how many they might be able to transfer into a new computer section. A survey is conducted and it is found that 20% of the 30 randomly selected employees have had some computer programming experience. Construct a confidence interval within which you are 95% sure the true proportion of the company's 40,000 employees with programming experience will lie.

TWO SAMPLE PROBLEMS:

18. Keypon Truckin' Company is trying to decide whether to purchase brand X or brand Y tires for its fleet of trucks. To estimate the difference in the two brands, an experiment is conducted using 10 of each brand. The tires are run until they wear out. The results are:

    Brand X: $\bar{X}_1$ = 23,200 miles, s_1 = 3,300 miles

    $$\Sigma(X_{1j} - \bar{X}_1)^2 = 98,010,000$$

    Brand Y: $\bar{X}_2$ = 24,100 miles, s_2 = 3,900 miles

    $$\Sigma(X_{2j} - \bar{X}_2)^2 = 136,890,000$$

    Assuming the populations are approximately normally distributed, determine the 95% confidence interval for $\mu_2 - \mu_1$.

19. A light bulb manufacturer wishes to compare the mean service life of two different special bulbs. Because a test of service life destroys the bulb, only a small random sample of each bulb is taken. The samples are from normal populations with equal variances. The sample results are as follows:

Sample	n	$\bar{X}_i$	s
1	6	1243 hrs.	152
2	4	985 hrs.	189

    (a) What are the best estimates for σ^2 and σ_d^2?
    (b) Find a 95% confidence interval for $\mu_1 - \mu_2$.

20. A variety store manager wishes to compare the mean time that customers spend in his store with the mean time spent in a competitor's store. Random samples of 18 and 10 customers respectively were selected. The sample results show that customers spend 30 minutes, on the average, in his store as opposed to 36 minutes in his competitor's store. The standard deviations were 16 minutes and 12 minutes respectively.

    (a) Construct a 90% confidence interval for the difference between the mean times for the two stores.
    (b) State the assumptions that are necessary for this problem.

21. A cigarette manufacturer tests tobaccos of two different brands for nicotine content and obtains the following results in milligrams:

   Brand X: 24, 26, 25, 22, 23, 24, 26, 28, 22, 25

   Brand Y: 27, 28, 25, 29, 26, 25, 28, 27, 28

   Construct a 95% confidence interval for $\mu_1 - \mu_2$.

22. A feedlot operator buys calves to raise himself. He buys from two different ranches, El Rancho BS and The DJC Stud Farm. The feedlot operator's records show that 30% of the calves bought from El Rancho BS have required care from a veterinarian, but only 25% of those from DJC Stud Farm have required such care. He has bought fifty calves from each ranch. Find an interval within which you can be 95% confident that the difference between proportions of "sick" animals from the two ranches lies.

23. A poll is taken among residents of a city and the surrounding county to determine their reaction to having the city and county governments merged. Of 5,000 city residents polled, 3,600 favor merger and 1,200 of 2,000 county residents polled favor it. Find the 95% confidence interval for the true difference in the percentages favoring the merger.

CHAPTER 8:   Solutions to Review Problems

ONE SAMPLE PROBLEMS:

1. **Solution**

The point estimate for μ is $\bar{X}$ = 7.5. We know that the true population standard deviation is 0.6; hence the confidence interval is $\bar{X} \pm Z_{\alpha/2} \cdot \sigma/\sqrt{n}$. Here $Z_{.025}$ is the value such that the area greater than it is .025; hence $Z_{.025}$ = 1.96. The confidence interval is

$$7.5 \pm 1.96 \cdot \frac{0.6}{\sqrt{36}} \text{ or } 7.5 \pm .196$$

I am 95% confident that the true mean amount of soft drink dispensed by this machine falls between 7.304 ounces and 7.696 ounces.

2. **Solution**

Here the sample standard deviation is given, but not the population standard deviation; therefore, the t-distribution is used. We are given $\bar{X}$ = 30.8, s = .63, n = 9.
For $(1-\alpha)$ = .98, $\alpha/2$ = .025 and degrees of freedom of n - 1 = 8, we have $t_{(.01,8)}$ = 2.896.
The 98% CI for μ:

$$\bar{X} \pm t_{(.01,8)} \cdot s/\sqrt{n}$$

$$30.8 \pm 2.896 \cdot \frac{.63}{\sqrt{9}} \text{ or } 30.8 \pm .60816.$$

I am 98% confident that the true mean weight of one quart of Lowlife Ice Cream is in the interval, 30.192 ounces to 31.408 ounces.

3. **Solution**

Here, $n = 5$, $\Sigma X = 416$, $\Sigma X^2 = 34{,}658$, hence $\bar{X} = \frac{416}{5} = 83.2$ and

$$s^2 = \frac{34658 - \frac{(416)^2}{5}}{4} = 11.7.$$

Since σ is unknown, use the t-table; $(1-\alpha) = .90$, so $\alpha/2 = .05$ and degrees of freedom, $n-1$, is 4. Hence, $t_{(.05,4)} = 2.132$.

90% CI for μ is $\bar{X} \pm t_{(.05,4)} \cdot s/\sqrt{n}$

$$83.2 \pm 2.132 \cdot \frac{\sqrt{11.7}}{\sqrt{5}} \text{ or } 83.2 \pm 3.261$$

The mean length of Froz Handsoff's jumps from the 90 meter hill falls in the interval 79.939 meters to 86.461 meters. I am 90% confident in this statement.

4. **Solution**

$$n = 41, \bar{X} = \frac{881.5}{41} = 21.5 \text{ meters};$$

$$s^2 = \frac{19873.85 - \frac{(881.5)^2}{41}}{40} = 23.04.$$

Confidence of 90% implies $\alpha/2 = .05$; since the true variance of the population is known, use the t-table: $t_{(.05,40)} = 1.684$. Therefore, the 90% CI for μ is

$$\bar{X} \pm t_{(.05,40)} \cdot \frac{s}{\sqrt{n}} = 21.5 \pm 1.684 \cdot \frac{\sqrt{23.04}}{\sqrt{41}}$$

or 21.5 ± 1.262.

I am 90% confident that the true mean distance that a typical soldier can throw a nuclear handgrenade is between 20.238 meters and 22.762 meters.

5. **Solution**

We are given grouped data.

V_i = class marks	f_i	f_iV_i	V_i^2	$f_iV_i^2$
25	10	250	625	6250
34	30	1020	1156	34680
43	25	1075	1849	46225
52	15	780	2704	40560
	n=80	Σf_iV_i=3125		$\Sigma f_iV_i^2$=127715

Therefore, $\bar{X} = \frac{3125}{80} = 39.0625$,

$$s^2 = \frac{127715 - \frac{(3125)^2}{80}}{79} = 71.45174$$

Since 79 degrees of freedom is not in the t-table, use the nearest: $t_{(.025,79)}$ is approximately $t_{(.025,60)} = 2.00$.
Therefore, 95% CI for μ is $\bar{X} \pm t_{(.025,79)} \cdot s/\sqrt{n}$

$$39.0625 \pm 2.00 \cdot \frac{\sqrt{71.45174}}{\sqrt{79}} = 39.0625 \pm 1.9021$$

I am 95% confident that the mean age of new car buyers in the midwest is in the interval 37.16 years to 40.96 years.

6. **Solution**

We are told $\sigma = 27.90$; also, n = 100, $\bar{X} = 157.45$, $1-\alpha = .92$. Therefore, $\alpha/2 = .04$ and $Z_{.04} = 1.75$. A 92% CI on μ is $\bar{X} \pm Z_{.04} \cdot \sigma/\sqrt{n}$,

$$157.45 \pm 1.75 \cdot \frac{27.90}{\sqrt{100}} = 157.45 \pm 4.8825$$

The true mean weekly wage rate for the secretaries falls between $152.5675 and $162.3325. I am 92% confident in this statement.

7. **Solution**

$n = 19$, $\bar{X} = 73{,}249$, $s = 37{,}246$, $(1-\alpha) = 90$; therefore, degrees of freedom are 18, $\alpha/2 = .05$ and $t_{(.05,18)} = 1.734$. A 90% CI is $\bar{X} \pm t_{(.05,18)} \cdot s/\sqrt{n}$.

$$73249 \pm 1.734 \cdot \frac{37246}{\sqrt{19}} = 73249 \pm 14816.7$$

The mean size of claims for fire damage in apartment complexes with 10 to 20 units falls between \$58,432 and \$88,066. I am 90% confident that this statement is correct.

8. **Solution**

$n = 100$, $X = 40$ successes, we want a 95% CI on p. We have two ways we could work this problem. The first involves using Table VII which provides confidence intervals based on exact binomial probabilities. The second method is based on the normal approximation to the binomial. Since Table VII provides only 95% confidence limits, it is beneficial to know both methods for use when other amounts of confidence are desired. Using Table VII the interval is .30% to .50%. The proportion of all citizens strongly against nuclear power as the primary source of future energy is between .30 and .50. I am 95% confident that this statement is correct.

9. **Solution**

$n = 100$, $X = 82$. We want a 95% CI on p. Enter Table VII with $n - x = 100 - 82 = 18$; read the limits as $100 - 27 = 73$ and $100 - 11 = 89$. The 95% confident limits for the proportion believing it is impossible to balance the budget are .73 and .89.

10. **Solution**

$n = 1200$, $X = 420$. Find 88% CI on p. Using the normal approximation, $f = 420/1200 = .35$, $Z_{.06} = 1.555$. The 88% CI for p is

$$f \pm Z_{.06} \cdot \sqrt{\frac{f(1-f)}{n}}$$

$$.35 \pm 1.555 \cdot \sqrt{\frac{.35 \cdot .65}{1200}} = .35 \pm .0214.$$

The 88% CI for the proportion of American made car owners that are satisfied with the maintenance service provided by dealerships is .3286 to .3714.

11. <u>Solution</u>

We want an interval that is $\pm$ \$5; hence, the width can be 10. We are given that $\sigma = 15$, so far $1-\alpha = .98$, $Z_{.01} = 2.326$ (from t-table with $\nu = \infty$). Using equation 8.23,

$$n = \frac{4 \cdot Z_{.01}^2 \cdot \sigma^2}{w^2} = \frac{4 \cdot 2.326^2 \cdot 15^2}{10^2} = 48.69$$

Therefore, use $n = 49$.

12. <u>Solution</u>

To be conservative, estimate σ with the larger value, \$300,000. 95% confidence implies $Z_{.025} = 1.96$. Hence,

$$n = \frac{4 \cdot Z_{.025}^2 \cdot \sigma^2}{w^2} = \frac{4 \cdot 1.96^2 \cdot 300,000^2}{50,000^2} = 553.19.$$

Therefore, rounding up, use $n = 554$.

13. <u>Solution</u>

We want the width of the 90% confidence interval to be .06 (i.e., $\pm$ 3%). Since $1-\alpha = .90$, $Z_{.05} = 1.645$ (from normal table). Using equation 8.37,

$$n = \frac{4 \cdot Z_{\alpha/2}^2 \, f(1-f)}{w^2} = \frac{4 \cdot 1.645^2 \cdot (.5) \cdot (.5)}{(.06)^2}$$

$$= \left(\frac{1.645}{.06}\right)^2 = 751.67$$

Therefore, use $n = 752$. Note: See table 8.4 in text for explanation for using $f = .5$.

14. **Solution**

We observed X = 9 blemished tires out of n = 120 tires. For 90% confidence, $Z_{.05} = 1.645$. The proportion in the sample that are blemished is $f = X/n = .075$. The 90% CI for p is $f \pm Z_{\alpha/2} \cdot \sqrt{f(1-f)/n}$

$$.075 \pm 1.645 \sqrt{\frac{(.075)(.925)}{120}} = .075 \pm .0396$$

The 90% confidence limits are .0354 and .1146.

15. **Solution**

n = 81, $\bar{X}$ = 2.7 seconds, s = 1.2 seconds. Since the true population variance is unknown, we must use the t-distribution. Since n-1 = 80 degrees is not listed, use the nearest or 60: $t_{(.05,80)} \doteq t_{(.05,60)} = 1.671$. Therefore, the CI is $\bar{X} \pm t_{(.05,80)} \cdot s/\sqrt{n}$:

$$2.7 \pm 1.671 \cdot \frac{1.2}{\sqrt{81}} = 2.7 \pm .2228$$

I am 90% sure that the true mean processing time for jobs on the new computer will be in the interval 2.48 seconds to 2.92 seconds.

16. **Solution**

n = 100 accounts, X = 22 with errors. Using Table VIII, the 95% confidence limits for the proportion of accounts with errors is .14 to .31.

17. **Solution**

n = 30, f = .20; hence, X = 6. Using Table VIII, the 95% confidence interval for the proportion with programming experience is .08 to .39.

TWO SAMPLE PROBLEMS:

18. **Solution**

We want a 95% confidence interval for the difference between two population means.

Sample	Size	Degrees of Freedom	Mean	s_i^2	Sum of Squares
1	10	9	23,200	3300^2	98,010,000
2	10	9	24,100	3900^2	136,890,000
	20	18			234,900,000

Pooled variance, $s^2 = \dfrac{\text{Pooled SS}}{\text{Pooled DF}} = \dfrac{234{,}900{,}000}{18}$

$= 13{,}050{,}000.$

By equation, 8.31, $s_d = \sqrt{s^2\left(\dfrac{1}{n_1}+\dfrac{1}{n_2}\right)} = 1615.5494$

For 95% confidence, $\alpha/2 = .025$, degrees of freedom = 18; $t_{(.025,18)} = 2.101$.
The 95% CI for $(\mu_2 - \mu_1)$ is
$(\bar{X}_2 - \bar{X}_1) \pm t_{(.025,18)} \cdot s_d$

$(24{,}100 - 23{,}200) \pm 2.101 \cdot 1615.5494$

900 ± 3394.27

I am 95% confident that the true difference in mileage between mean of Brand Y tires and the mean of Brand X tires falls in the interval

-2494.27 miles to 4294.27 miles.

19. <u>Solution</u>

Sample	n	Degrees of Freedom	$\bar{X}_i$	s_i^2	Sum of Squares $(n_i-1)\cdot s_i^2$
1	6	5	1243	152^2	115,520
2	4	3	985	189^2	107,163
	10	8			222,683

a. The best estimate of σ^2 is the pooled variance,

$s^2 = \dfrac{\text{Pooled SS}}{\text{Pooled DF}} = \dfrac{222{,}683}{8} = 27{,}835.375$

The best estimate of σ_d is, using equation 8.31,

$s_d = \sqrt{s^2\left(\dfrac{1}{n_1}+\dfrac{1}{n_2}\right)} = 107.6943$

b. For 95% confidence, $\alpha/2 = .025$, pooled degrees of freedom = 8; hence, $t_{(.025,8)} = 2.306$. Therefore, $(\bar{X}_1 - \bar{X}_2) \pm t_{(.025,8)} \cdot s_d$ is

$$(1243-985) \pm 2.306 \cdot 107.6943$$
$$258 \pm 248.3$$

I am 95% confident the true difference in service life for the two populations of bulbs, $\mu_1 - \mu_2$, falls in the interval 9.7 hrs to 506.3 hrs.

20. **Solution**

   His store: $n_1 = 18$, $\bar{X}_1 = 30$ minutes, $s_1 = 16$

   Competitor's store: $n_2 = 10$, $\bar{X}_2 = 36$ minutes, $s_2 = 12$

   a. 90% CI for $(\mu_1 - \mu_2)$: $(\bar{X}_1 - \bar{X}_2) \pm t_{(.05,26)} \cdot s_d$.

   $t_{(.05,26)} = 1.706$.

   $$s^2 = \frac{(n_1-1) \cdot s_1^2 + (n_2-1) s_2^2}{n_1 + n_2 - 2} = \frac{4352 + 1296}{26} = 217.2308$$

   $$s_d = \sqrt{s^2 \left(\frac{1}{n_1} + \frac{1}{n_2}\right)} = 5.813$$

   The 90% CI for $(\mu_1 - \mu_2)$ is $-6 \pm 1.706 \cdot 5.813 = -6 \pm 9.917$ or -15.917 minutes to 3.917 minutes.

   b. Both populations must be assumed to have normal distributions with their true variances equal. The samples must be independent at each other.

21. **Solution**

   Here raw data is given. Summarize data for each sample.

Sample	n_i	n_i-1	ΣX	ΣX^2	$\bar{X}_i$	$\Sigma X^2 - \frac{(\Sigma X)^2}{n}$
1	10	9	245	6035	24.5	32.5
2	9	8	243	6577	27.0	16.0
	19	17			Pooled SS =	48.5

$$s^2 = \frac{48.5}{17} = 2.85294; \quad s_d = \sqrt{s^2\left(\frac{1}{n_1} + \frac{1}{n_2}\right)} = .7761$$

95% CI for $(\mu_x - \mu_y)$: $(\bar{X}_1 - \bar{X}_2) \pm t_{(.025,17)} \cdot s_d$

$-2.5 \pm 2.110 \cdot .7761 = -2.5 \pm 1.6375$

The 95% confidence interval for the true difference of mean nicotine content for two brands of cigarettes is -4.1375 to -.8625 milligrams.

22. <u>Solution</u>

We want a 95% confidence interval for the difference of two population proportions.

Sample	f_i	n_i	$f_i(1-f_i)/n_i$
1. El Rancho BS	.30	50	.00420
2. DJC Stud Farm	.25	50	.00375
			.00795

Hence, $s_{f_1-f_2} = \sqrt{\frac{f_1(1-f_1)}{n_1} + \frac{f_2(1-f_2)}{n_2}} = \sqrt{.00795} = .08916$

The 95% CI for $(p_1 - p_2)$ is

$(f_1-f_2) \pm Z_{.025} \cdot s_{f_1-f_2} = .05 \pm 1.96 \cdot .08916$

$= .05 \pm .17$

I am 95% confident that the true difference of the proportions of calves needing care is in the interval -.12 to .22.

23. <u>Solution</u>

X_i counts the number that favor the merger in sample i.

Sample	n_i	X_i	f_i	$f_i(1-f_i)/n_i$
1. City	5000	3600	.72	.00004032
2. County	2000	1200	.60	.00012000
				.00016032

$$s_{f_1-f_2} = \sqrt{\frac{f_1(1-f_1)}{n_1} + \frac{f_2(1-f_2)}{n_2}} = \sqrt{.00016032} = .01266$$

The 95% confidence interval for $p_1 - p_2$ is

$$(f_1 - f_2) \pm Z_{\alpha/2} \cdot s_{f_1-f_2}$$

$(.72-.60) \pm 1.96 \cdot .01266 = .12 \pm .0248.$

I am 95% confident that the true difference between the proportion of city and county residents that favor the merger is in the interval .0952 to .1448.

Tests of Hypotheses

Major Topics and Key Concepts

    Objective of hypothesis testing: Deciding whether or not a statement about a population parameter is correct and controlling the probability of making the wrong decision.

9.1 Establishing Hypotheses

- Null hypothesis, H_0: a statement concerning what the parameter value is supposed to be, or is claimed to be, or has been in the past.
- Alternative hypothesis, H_a: a statement concerning values of the parameter that would be of interest to the investigator if the null hypothesis is rejected.

9.2 Testing Hypotheses

9.3 Type I and Type II Errors

- Type I error: in reality H_0 is true, but the evidence in the sample indicates that it should be rejected.
- α = P(Type I error) is called the "level of significance" of the test.
- Type II error: in reality H_0 is false, but the evidence in the sample does not indicate that it should be rejected.
- β = P(Type II error)
- $1 - \beta$ = Power of the test; the probability of rejecting H_0 when H_0 should be rejected.

9.4 The Steps of Hypothesis Testing

- State the alternative hypothesis such that, if H_0 is rejected, H_a contains values of the parameter that would be of particular interest to the investigator, or values that would lead him to take action. (Never put the equality sign in H_a.)
- Rejection region contains the values of the statistic that support the rejection of H_0.

- Possible Decisions: We either reject or we do not reject H_0.
    1. If H_0 is rejected, then claim H_a is true. Risk of this decision being wrong is α which is known.
    2. If H_0 is not rejected, then claim there is insufficient evidence in the data to indicate H_0 is false. Do not claim H_0 is true, since the risk of that decision being wrong, β, is usually known.

ONE SAMPLE TESTS:

9.5 Hypotheses on a Normal Mean

- Correct statistic depends upon whether σ is known.
- Rejection region is determined by using H_A, the α-level and either the normal or t-tables.

9.6 Hypotheses on a Binomial p

- Test statistic and rejection region are based on the normal approximation.

9.7 A Connection Between Testing Hypotheses and Estimation

TWO SAMPLE TESTS:

9.8 Testing the Hypothesis $\mu_1 = \mu_2$

- Based on Sections 7.6 and 8.7.
- Used for comparing two <u>independent</u> samples.

9.9 Paired Comparisons

- Used for comparing two <u>dependent</u> samples; i.e., when there is some <u>relationship</u> between a pair of experimental units.
- Reduces to a one sample test similar to Section 9.5.

9.10 Group Versus Pairs

9.11 Testing the Hypothesis that $p_1 = p_2$

- Based on Sections 7.7 and 8.8.

9.12 Testing the Hypothesis that $\sigma_1^2 = \sigma_2^2$ and the F Distribution

- Used only for independent samples.

- $F = s^2_{larger}/s^2_{smaller}$
- Use Table VIII for rejection regions.

9.13 Summary

- Table 9.1 aids selection of appropriate test statistic.

CHAPTER 9:   Review Test

1. The null hypothesis is established in such a way that it states nothing is different than it is supposed to be, or it is claimed to be, or it has been in the past.
   (a) True.
   (b) False.

2. If strong evidence in the sample suggests that the null hypothesis is incorrect:
   (a) Then hypothesis testing cannot be performed.
   (b) We reject it and accept the alternative hypothesis.
   (c) We formulate a new one and repeat the testing process.
   (d) It becomes an alternative hypothesis.

3. A set R of $\bar{X}$-values that will lead to a rejection of H_o and a preference for H_a is _____
   (a) The test parameters.
   (b) A rejection region.
   (c) A critical region.
   (d) (b) or (c)
   (e) (a) or (c)

4. Not rejecting the null hypothesis when it is actually false is:
   (a) Not possible if the test is performed properly.
   (b) A Type I error.
   (c) A Type II error.
   (d) Is identified by the Greek letter α.

5. The size of a test or its level of significance refers to:
   (a) Its probability α of a Type I error.
   (b) Its power, $1-\beta$.
   (c) Its probability β of a Type II error.
   (d) The sample size.

6. Which of the following statements about α and β is not correct?
   (a) α and β do not ordinarily sum to 1.
   (b) If $\alpha = 0$, then we must always accept H_o.
   (c) α and β are inversely related.
   (d) α and β may be arranged so that they are both equal to zero.

7. When problems are solved by hypothesis testing:
    (a) The null and alternative hypotheses must be formulated in terms of parameter values.
    (b) A level of significance and sample size should be set.
    (c) A rejection region should be specified.
    (d) All of the above.
    (e) (a) and (c) only.

8. The form of the alternative hypothesis determines the rejection region selected.
    (a) True.
    (b) False.

9. When the alternative hypothesis is of the "not equal to" type, then the rejection region _____
    (a) Is smaller.
    (b) Occurs in only one tail.
    (c) Occurs in both tails.
    (d) Is larger.

10. Chosing a small value of α to guard against a Type I error, will at the same time insure that β will also be small.
    (a) True.
    (b) False.

11. Which of the following statements could never be an alternative hypothesis?
    (a) $H_a: \mu \leq 10$.
    (b) $H_a: \mu < 10$.
    (c) $H_a: \mu \neq 10$.
    (d) $H_a: \mu > 10$.

12. The rejection region consists of values of the test statistic and should be selected in such way that:
    (a) The probability of the statistic falling in the region is α if H_o is true.
    (b) The values of the parameter in the region support the alternative hypothesis, H_a.
    (c) The boundary point or points of the region are determined from the table corresponding to the distribution of the test statistic.
    (d) All of the above.

13. The rejection region always consists of values of the statistic in either one or both tails and never contains the point t = 0.
    (a) True
    (b) False

14. Whenever we use the t distribution for calculating confidence intervals or for testing hypotheses, we are assuming that the data _____
    (a) Form a random sample.
    (b) Are from a normal population.
    (c) If the population is not normal, the sample is large enough that the central limit theorem holds.
    (d) All of the above.
    (e) (a) and (b).

15. In testing a hypothesis on a binomial p, the standard deviation of X must be estimated from a sample.
    (a) True.
    (b) False.

16. Given a level-α test, we can construct a $100(1-\alpha)\%$ confidence interval. The interval would include:
    (a) All the null hypotheses not rejected by the test.
    (b) All the alternative hypotheses not accepted by the test.
    (c) All the null hypotheses rejected by the test.
    (d) None of the above.

17. For a one-tailed test with $\begin{cases} H_o: \mu_1 = \mu_2 \\ H_a: \mu_1 < \mu_2 \end{cases}$ we use _____ tail for the critical region.
    (a) The upper
    (b) The lower
    (c) Both upper and lower
    (d) None of the above

18. The t-statistic for the text in Problem 17 is
    (a) $\dfrac{\bar{X}_1 - \bar{X}_2}{s_d}$
    (c) $\dfrac{\bar{X}_1 - \bar{X}_2}{s^2}$
    (b) $\dfrac{\bar{X}_2 - \bar{X}_1}{s^2}$
    (d) $\dfrac{\bar{X}_2 - \bar{X}_1}{s_d}$

19. Which of the following assumptions, if any, may be relaxed in comparing or testing differences between two samples?

    (a) Randomness.
    (b) Normality.
    (c) Common variance.
    (d) None of the above.

20. Paired comparisons can increase the precision of an experiment if

    (a) The variation within the pairs is less than the variation among pairs.
    (b) The between treatment variation is less than the within treatment variation.
    (c) None of the basic assumptions are relaxed.
    (d) Less homogeneous experimental material is used.

21. When paired comparisons are used as opposed to group comparisons, there is _____ degrees of freedom.

    (a) A gain in
    (b) A doubling of
    (c) No change in the number of
    (d) A loss of

22. If group analysis techniques are used on data from a paired experiment, the estimate of variance will be:

    (a) Unaffected.
    (b) Twice as large as the estimate from paired analysis.
    (c) Too large.
    (d) Too small.

23. When data from a group experiment are paired, the results may include:

    (a) A variance estimate that is too large.
    (b) A variance estimate that is too small.
    (c) A loss in degrees of freedom.
    (d) All of the above.

24. With two samples from binomial populations are used to compare two proportions, the best estimate for p will be obtained by pooling the two samples with number of successes weighted by sample size.

    (a) True.
    (b) False.

25. The F-test allows one to
    (a) Determine if the variances of any two normal populations are equal.
    (b) Check the validity of the assumption of equal variances which is needed for testing two independent population means.
    (c) Determine if two population means are equal.
    (d) Both (a) and (b).
    (e) All of the above.

26. To test $\begin{cases} H_o: \sigma_1^2 = \sigma_2^2 \\ H_a: \sigma_1^2 \neq \sigma_2^2 \end{cases}$ if s_2^2 is larger in value than s_1^2 then the degrees of freedom for the F-table are
    (a) $\nu_1 = n_1-1$, $\nu_2 = n_2-1$
    (b) $\nu_1 = n_2-1$, $\nu_2 = n_1-1$
    (c) $\nu_1 = n_1-1$, $\nu_2 = n_1+n_2-2$
    (d) $\nu_1 = 2$, $\nu_2 = n_1+n_2-2$

CHAPTER 9: Answers to Review Test

Question	Answer	Text Section Reference
1	a	9.1
2	b	9.1
3	d	9.2
4	c	9.3
5	a	9.3
6	d	9.3
7	d	9.4
8	a	9.4
9	c	9.4
10	b	9.4
11	a	9.5
12	d	9.5
13	a	9.5
14	d	9.5 (after third example)
15	b	9.6 (use p_o in the standard deviation)
16	a	9.7
17	$\begin{bmatrix} b \\ a \end{bmatrix}$ or $\begin{bmatrix} a \\ d \end{bmatrix}$ (either combination is correct)	9.8
18		9.8
19	b	9.8
20	a	9.9
21	d	9.10
22	c	9.10
23	d	9.10
24	a	9.11
25	d	9.12
26	b (since larger sample variance is in numerator)	9.12

CHAPTER 9: Review Problems

ONE SAMPLE PROBLEMS:

1. A high intensity bulb for home movie projectors should have a mean life of 25 hours. For each situation state the appropriate null and alternative hypotheses that should be tested.
    (a) A consumer protection service will publically accuse the company of fraudulent claims if their evidence indicates that the true life is not 25 hours.
    (b) The company will guarantee the bulb for 25 hours if its evidence indicates that the life exceeds this.
    (c) A quality control engineer wants to know if the bulb life differs from laboratory specifications.

2. An oil company will build a new service station on a given corner lot if an average of more than 2,000 cars per day pass the lot. It is assumed that the universe standard deviation is 100 cars per day. If a sample of 36 days resulted in a mean of 2020 cars per day, a standard deviation of 92 cars per day, what should the oil company do if they are willing to take a 10% risk of making a Type I error?

3. An electronic tube manufacturer claims that his product has a mean service life of 250 hours or more. A firm that requires a larger number of these tubes wants to test the claim. If the claim is correct, they will purchase the product; otherwise, they will seek another supplier.
    (a) What would the appropriate H_o and H_a be for this problem?
    (b) What are the Type I and Type II errors associated with this problem?
    (c) Assume a random sample of 25 tubes is taken and the mean life is found to be 225 hours with a standard deviation of 50 hours. If a 5% level of significance is selected, what should the company do?

4. A shipment of several thousand parts should have a mean thickness of 32 mm. If the evidence in a sample of 49 parts indicates that this is not so, then the entire shipment will be sent back. The probability of a Type I error is chosen to be .05.
    (a) State the hypotheses.

(b) State the rejection region.
(c) If $\bar{X}$ = 31.94 mm and s^2 = .0441, what is your conclusion?

5. It has been shown that the fruit grown near a certain type of manufacturing plant absorbs lead which is toxic to humans. When the level of lead is less than 32 parts per million (ppm) in the fruit, the fruit is not considered to be toxic. The Food and Drug Administration periodically tests the fruit in an orchard. They assume the fruit is toxic unless there is strong evidence to the contrary. The toxicity level of lead in the fruit has been shown to be normally distributed about its mean. If the FDA sampled 9 trees and found a mean of 25 ppm and a standard deviation of 5 ppm, what action should be taken if alpha is .10?

6. Former football wide receiver L. G. Dunn now runs a bar in Houston called "L. G.'s Down and Out." Past records indicate that he only sells a mean of 180 gallons of beer per night. An advisor from the Small Business Administration suggests that adding a game room would increase sales. He adds four pinball machines and plans to construct an entire new wing if sales warrant it. Using an alpha of .10, what action should be taken if after a sample of 25 days the mean sales were 165 gallons and s = 12?

7. An automobile dealer believes that .80 of his 1981 cars were sold with radios installed. Test this belief with an alpha of .12. Assume you were able to find that in a sample of 144 cars 108 of them were sold with radios. What conclusion would you reach?

8. A manufacturer claims that at least 95 percent of the equipment which he supplies to factories conforms to specifications. An examination of 200 randomly selected equipment items reveals that 18 are faulty. Test the manufacturer's claim at a significance level of .01. Is he justified in making his claim on the basis of the sample results?

9. If the percentage of automobile owners that has no confidence in the maintenance service provided by dealerships is more than 60%, then a major corporation will initiate an extensive advertising campaign in order to bolster confidence in "Mr. Goodwheels." If 378 owners surveyed from a sample of 600 have no confidence in the dealership service, then using α = .06, what conclusion should be reached.

TWO SAMPLE PROBLEMS:

10. An automatic paint machine is designed to spray paint an automobile with an average of 40 ounces of paint. The data below show the amount of paint used to spray a sample of automobiles on two successive days.

    Day 1: 38, 42, 35, 40, 48, 33, 44, 50, 39
    Day 2: 43, 39, 45, 52, 46, 30

    (a) Determine if the machine used a different mean amount of paint on the two days, using alpha of .05.
    (b) State all assumptions necessary to use this test method.

11. A company uses two different methods of training management prospects. Method 1 is the traditional approach; method 2 is a new, more costly program. The new method will be implemented on a permanent basis if it is better. New trainees were randomly assigned to one of the two methods and graded on performance 3 months after completion of the program. Use $\alpha = .10$.

	Method 1	Method 2
Mean score	68.0	72.0
Score standard deviation	3.4	3.8
Number of trainees	20.	16.

12. A company wished to study the effectiveness of a coffee break on the productivity of its workers. It selected 8 workers and measured their productivity on a day without a coffee break, and later measured their productivity on a day when they were given a coffee break. The scores measuring the productivity were numbers of units completed in a shift, adjusted for defective units or those needing reworking. The scores were as follows:

Worker	1	2	3	4	5	6	7	8
No Coffee break	24	34	30	32	44	31	33	35
Coffee break	29	37	32	36	43	29	30	39

    (a) What method of testing should be used?
    (b) Do these scores indicate that a coffee break raises productivity for $\alpha = .05$?

13. Skinny Magazine recently conducted a poll of CIA and FBI agents concerning the use of unauthorized wiretaps to gain information about possible government subversives. The magazine wanted to determine if the proportion of FBI agents that use illegal wiretaps is greater than the proportion of CIA agents. Use .025

level of significance. The results of the Skinny poll are as follows:

Agency	Sample Size	Number That Have Used Wiretaps
1. FBI	40	24
2. CIA	60	24

14. In a study to estimate the proportion of housewives who own automatic dishwashers, it was found that 60 of 150 urban residents had a dishwasher while only 35 of 125 rural residents owned a dishwasher. Determine if the proportion of all rural homes that own dishwashers is different than that of urban homes at the 8% level of significance.

15. Test the hypothesis of equal variances for Problem 10 using the .10 level of significance.

16. Test the hypothesis of equal variances for Problem 11 using the .05 level of significance.

CHAPTER 9: Solutions to Review Problems

ONE SAMPLE PROBLEMS:

1. Solution

   The correct set of hypotheses depends upon the reasons for which the experiment is conducted:

   a. H_o: $\mu \geq 25$ hours ⟹ the claim is true.
      H_a: $\mu < 25$ hours ⟹ the claim is false; take action against the company.

   b. H_o: $\mu \leq 25$ hours ⟹ life is not long enough to justify the guarantee.
      H_a: $\mu > 25$ hours ⟹ true life does exceed 25 hours; the guarantee can be given.

   c. H_o: $\mu = 25$ hours ⟹ the production process is in control.
      H_a: $\mu \neq 25$ hours ⟹ bulb life does differ from lab specifications; adjustments in production process are necessary.

2. Solution

   Follow the hypothesis testing steps.
   Steps:
   1&2. H_o: $\mu \leq 2000$ cars ⟹ do not build the station.
        H_a: $\mu > 2000$ cars ⟹ build a new station.
   3. $\alpha = .10$, $n = 36$.
   4. Here, we know the true population standard deviation, $\sigma = 100$; and, we also have calculated the sample standard deviation, $s = 92$. If both are

available, always use σ. Therefore, the appropriate test statistic is

$$Z = \frac{\bar{X} - \mu_0}{\sigma/\sqrt{n}}$$

or, R: $Z > 1.282$.

The tail of the rejection region is determined from H_a; the area under the curve in the tail is α; the critical value for this problem comes from the normal distribution (Table V) or, equivalently, from the t-distribution with <u>infinite</u> degrees of freedom (Table VI, last row).

5. $Z = \dfrac{\bar{X} - \mu_0}{\sigma/\sqrt{n}} = \dfrac{2020-2000}{100/6} = 1.2$

6. Since the calculated $Z = 1.2$ does not fall in the rejection region, the decision is that H_0 should not be rejected. Therefore, this sample does not provide sufficient information to warrant the construction of a new service station.

3. <u>Solution</u>

μ = true mean service life for this type of tube.

a. H_0: $\mu \geq 250$ hours ⇒ purchase product.

H_a: $\mu < 250$ hours ⇒ seek another supplier.

b. Type I error: Seeking a new supplier when in reality this manufacturer's product was good.
Type II error: Buying from this manufacturer when in reality the mean life was less than 250 hours.

c. $\alpha = .05$, $n = 25$; the population standard deviation is unknown; therefore, a t-statistic is used. The rejection region is in the lower tail because of H_a:

R: $t < -t_{(.05, 24)} = -1.711$

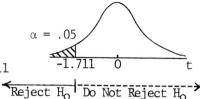

We now compare the sample $\bar{X}$ to the hypothesized population mean, μ_0:

$$t = \frac{\bar{X} - \mu_0}{S/\sqrt{n}} = \frac{225 - 250}{50/\sqrt{25}} = -2.5$$

Since the calculated $t = -2.5$ falls in the rejection region, reject H_0. Therefore, the true mean life of the electronic tube is less than 250 hours; the firm should seek a new supplier.

4. **Solution**

μ = true mean thickness of all parts in the shipment.

a. H_0: $\mu = 32$ mm ⇒ parts are good, keep them.

H_a: $\mu \neq 32$ mm ⇒ parts do not meet specifications, return to sender.

b. The population standard deviation is unknown; hence, the t-distribution is needed. $\alpha = .05$, $n = 49$, $t_{(.025, 48)} \doteq 2.021$.
R: $t < -2.021$ or $t > 2.021$.

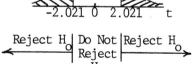

c. $t = \dfrac{\bar{X} - \mu_0}{S/\sqrt{n}} = \dfrac{31.94-32}{.21/7} = -2.0$.

Since $t = -2.0$ is not in the rejection region, do not reject H_0. Therefore, there is no evidence in this sample to indicate that the shipment should be returned.

5. Solution

μ = true mean level of lead in the fruit in orchard. The key here is that "they assume fruit toxic unless there is strong evidence to the contrary.

H_0: $\mu \geq 32$ ppm $\Rightarrow$ fruit is toxic.

H_a: $\mu < 32$ ppm $\Rightarrow$ fruit is not toxic.

Since the population standard deviation is unknown, use a t-statistic. With $\alpha = .10$, $n = 9$, then $t_{(.10,8)} = 1.397$.
R: $t < -1.397$

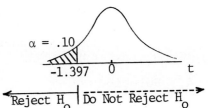

We now compare the sample mean, $\bar{X}$, to the hypothesized population mean, μ_0:

$t = \dfrac{\bar{X} - \mu_0}{S/\sqrt{n}} = \dfrac{25-32}{5/\sqrt{9}} = -4.2$.

Because the calculated $t = -4.2$ is in R, we reject H_0. Therefore, there is strong evidence to indicate that the fruit is <u>not</u> toxic.

6. Solution

μ = true mean gallons of beers sold per night with game room.

H_0: $\mu \leq 180$ $\Rightarrow$ game room is not effective.

H_a: $\mu > 180$ $\Rightarrow$ game room increases sales.

α = .10, n = 25,
σ is unknown.
$t_{(.10,24)}$ = 1.318;
hence, R: t > 1.318.

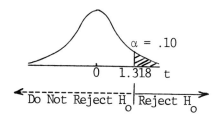

Now, comparing the sample mean to the hypothesized population mean, μ_o:

$$t = \frac{\bar{X} - \mu_o}{S/\sqrt{n}} = \frac{165-180}{12/\sqrt{25}} = -6.25$$

Since the calculated t = -6.25 is not in the rejection region, do not reject H_o. Therefore, there is no evidence that adding pinball machines increases sales for LG's Down and Out. (Note: do not let the sample data influence your choice of the hypotheses.)

7. Solution

This problem concerns a proportion rather than a mean. For each item in the sample we determine if it has the trait of interest or not (in this case, a radio). The random variable, r, is a <u>count</u> of the items in the sample with the trait. p = proportion of population of all 1981 cars that have a radio.

H_o: p = .80 the "claimed" proportion.

H_A: p ≠ .80 the true proportion is either more or less.

(Suppose the belief had been, or implied that, at least .80 have radios; than the alternative would be H_A: p < .80.)

When the normal approximation is used, always use normal table to determine the critical point; α = .12 and we have a two tailed test; hence, $Z_{.06}$ = 1.555, and R: Z < -1.555 or Z > 1.555.

Now from the sample, we have

r = 108 cars with radios in the sample
n = 144 cars that were sampled
f = r/n = .75 is the proportion of cars in the sample with radios.

We now compare f to p_0.

$$Z = \frac{f - p_0}{\sqrt{\frac{p_0(1-p_0)}{n}}} = \frac{.75 - .80}{.03333} = -1.50$$

Therefore, do not reject H_0; there is no evidence to indicate that the dealers belief is wrong.

8. Solution

p = true <u>proportion</u> of all equipment that the manufacturers supplies <u>that</u> <u>conforms</u> to specifications.

H_0: p ≥ .95
H_a: p < .95
α = .01, one tailed test; thus,
$Z_{.01}$ = 2.326, and
R: Z < -2.326.

From the sample, 18 of 200 are faulty; hence, 182 of 200 conform. Remember, p = proportion that conform. Now, f = 182/200 = .91.

$$Z = \frac{f - p_0}{\sqrt{\frac{p_0(1-p_0)}{n}}} = \frac{.86 - .95}{\sqrt{\frac{(.95)(.05)}{200}}} = -2.596$$

The calculated Z = -2.596 is in the rejection region. Conclude that the true proportion of the equipment that this manufacturer supplies to factories that conforms to specifications is actually less than .95. The test was conducted at the .01 level of significance.

9. Solution

p = proportion of owners with no confidence in maintenance service.

H_o: $p \leq .60$
H_a: $p > .60$ ⇒ initiate advertising campaign
$\alpha = .06$, one tail;
$Z_{.06} = 1.555$, thus
R: $Z > 1.555$

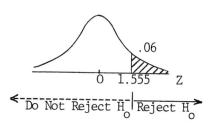

From the sample
r = 378 owners with no confidence
n = 600 sampled.
f = .63 proportion in sample with no confidence.

Comparing the sample proportion, f, to the hypothesized population proportion, p_o,

$$Z = \frac{f - p_o}{\sqrt{\frac{p_o(1-p_o)}{n}}} = \frac{.63 - .60}{\sqrt{\frac{(.60)(.40)}{600}}} = 1.5$$

The calculated Z = 1.5 is not in R; hence, do not reject H_o. There is not sufficient evidence in this sample to indicate that the true percentage that have no confidence in dealership maintenance service exceeds 60%.

TWO SAMPLE PROBLEMS:

10. <u>Solution</u>

Let μ_1 = true mean amount of paint used per car for all cars painted on day 1; let μ_2 be the same for day 2.

a. H_o: $\mu_1 = \mu_2$ ⇒ the same mean amount for the two days.

H_a: $\mu_1 \neq \mu_2$ ⇒ different mean amount used on the two days.

We have two independent samples with neither population variance known; hence, t-statistic is needed. The degrees of freedom are $n_1 + n_2 - 2 = 13$.

With $\alpha = .05$,
$t_{(.025, 13)} = 2.160$
R: $t < -2.160$ or $t > 2.160$.

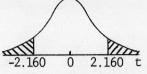

The statistic is

$$t = \frac{(\bar{X}_1 - \bar{X}_2) - (\mu_1 - \mu_2)_0}{S\sqrt{\frac{1}{n_1} + \frac{1}{n_2}}}.$$

The value $(\mu_1 - \mu_2)_0$ is what the hypothesized difference is in H_0; here this is zero. From the sample, we have

Day	n	df	ΣX	$\bar{X}$	ΣX^2	$\Sigma X^2 - \frac{(\Sigma X)^2}{n}$
1	9	8	369	41.0	15383	254.0
2	6	5	255	42.5	11115	277.5
	15	13		-1.5	Pooled SS =	531.5

$$s^2 = \frac{\text{pooled SS}}{\text{pooled df}} = \frac{531.5}{13} = 40.8846$$

$$S = \sqrt{40.8846} = 6.3941$$

Therefore, $t = \dfrac{(41.0-42.5) - 0}{6.3941\sqrt{\frac{1}{9} + \frac{1}{6}}} = \dfrac{-1.5}{3.370} = -.4451$

Since $t = -.4451$ is not in R, we do not reject H_0. There is no evidence to indicate that the mean amount of paint used per car was different for these two days.

b. The assumptions needed for this test procedure are:
1. The samples from the two days must be independent.
2. Amount of paint used per car must follow a normal distribution for both days; i.e., each day constitutes a population.
3. The true population variances must be equal for the two days.

11. <u>Solution</u>

Let μ_1 = true mean for all trainees receiving method 1 (old).

Let μ_2 = true mean for all trainees receiving method 2 (new).

H_o: $\mu_1 \geq \mu_2$

H_a: $\mu_1 < \mu_2$ ⟹ new method has greater mean.

Since $\alpha = .10$, the samples are independent, and the degrees of freedom are $n_1 + n_2 - 2 = 34$; then $t_{(.10, 34)} \doteq 1.310$.

R: $t < -1.310$

Comparing the observed difference of sample means to the hypothesized difference between population means, zero, the test statistic is

$$t = \frac{(\bar{X}_1 - \bar{X}_2)}{s\sqrt{\frac{1}{n_1} + \frac{1}{n_2}}}$$

Recall, s is the pooled estimate of common standard deviations. Since we are given two sample deviations, we need

$$s = \sqrt{\frac{(n_1-1)s_1^2 + (n_2-1)s_2^2}{n_1 + n_2 - 2}} = \sqrt{\frac{19 \cdot 3.4^2 + 15 \cdot 3.8^2}{34}}$$

$$= 3.5820$$

Therefore, $t = \dfrac{68-72}{3.5820\sqrt{\frac{1}{20} + \frac{1}{16}}} = -3.329$

Since the calculated $t = -3.329$ is in the rejection region, we reject H_o and conclude that the population of trainees receiving the new method do have higher mean performance score than the population of those receiving the traditional method.

12. __Solution__

   a. This experiment is set up as paired comparisons because the <u>same individual</u> is measured under both conditions.

   b. μ_1 = mean productivity of all workers with no coffee break.

   μ_2 = mean productivity of all workers having a coffee break; $\mu_D = \mu_1 - \mu_2$.

   $H_o: \mu_1 \geq \mu_2$ or $H_o: \mu_D \geq 0$ $\Rightarrow$ coffee break has greater mean productivity.
   $H_a: \mu_1 < \mu_2$ $H_a: \mu_D < 0$

   Since $\alpha = .05$, $n = 8$ paired observations, then $t_{(.05,7)} = 1.895$; hence,

   R: $t < -1.895$

   From the sample, calculate the difference for each individual; $D_j = X_{1j} - X_{2j}$,

   D_i: $-5, -3, -2, -4, +1, +2, +3, -4$.

   $\Sigma D_i = -12$, $\Sigma D_i^2 = 84$

   $\bar{D} = -1.5$, $s_D^2 = \dfrac{84 - \dfrac{(-12)^2}{8}}{7} = 9.4286$

   $s_D = 3.0706$, $s_{\bar{D}} = s_D/\sqrt{n} = 1.0856$

   Therefore,

   $$t = \dfrac{\bar{D} - \mu_o}{s_{\bar{D}}} = \dfrac{\bar{D} - \mu_o}{s_D/\sqrt{n}} = \dfrac{-1.5}{1.0856} = -1.382$$

   Do not reject H_o; there is insufficient evidence to indicate that coffee breaks increase productivity.

13. __Solution__

   This is test comparing two population proportions.

   p_1 = proportion of all FBI agents that use illegal wiretaps.

p_2 = proportion of all CIA agents that use illegal wiretaps.

H_o: $p_1 \leq p_2$
H_a: $p_1 > p_2$ ⇒ proportion is higher for FBI agents

$\alpha = .025$, one tailed test; therefore, $Z_\alpha = 1.96$ and
R: $Z > 1.96$.

$f_1 = r_1/n_1 = .60$
$f_2 = r_2/n_2 = .40$
$f_1 - f_2 = .20$

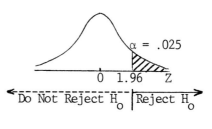

$f = \dfrac{r_1 + r_2}{n_1 + n_2} = \dfrac{80}{100} = .48$ is the proportion in the combined samples using illegal wiretaps.

Hence,

$$Z = \dfrac{f_1 - f_2}{\sqrt{f(1-f)(\frac{1}{n_1} + \frac{1}{n_2})}} = \dfrac{.20}{\sqrt{(.48)(.52)(\frac{1}{60} + \frac{1}{40})}}$$

$= 1.961$

Since the calculated $Z = 1.961$ is in the rejection region, we reject H_o. The magazine should conclude that a higher proportion of FBI agents use illegal wiretaps than CIA agents.

14. <u>Solution</u>

p_1 = proportion of all urban homes with dishwashers.

p_2 = proportion of all rural homes with dishwashers.

H_o: $p_1 = p_2$
H_a: $p_1 \neq p_2$
$\alpha = .08$
$Z_{.04} = 1.75$
R: $Z < -1.75$ or $Z > 1.75$

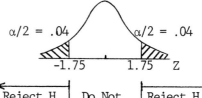

$$r_1 = 60 \qquad n_1 = 150 \qquad f_1 = 60/150 = .40$$
$$r_2 = 35 \qquad n_2 = 125 \qquad f_2 = 35/125 = .28$$
$$r_1 + r_2 = 95 \quad n_1 + n_2 = 275 \quad f = \frac{r_1 + r_2}{n_1 + n_2} = .34545$$

Therefore,

$$Z = \frac{(f_1 - f_2)}{\sqrt{f(1-f)(\frac{1}{n_1} + \frac{1}{n_2})}} = \frac{.40 - .28}{\sqrt{(.34545)(.65455)(\frac{1}{150} + \frac{1}{125})}}$$

$$= \frac{.12}{.05759} = 2.08$$

Since the calculated Z = 2.08 is in R, we reject H_o. We conclude that there is a difference in the proportion of rural and urban homes with dishwashers.

15. <u>Solution</u>

H_o: $\sigma_1^2 = \sigma_2^2$

H_a: $\sigma_1^2 \neq \sigma_2^2$ We will use $\alpha/2$ for F-table.

$\alpha = .10$, $n_1 = 9$, $n_2 = 6$. Before we can determine the rejection region, we need to know which sample variance is larger.

$$s_1^2 = \frac{\Sigma X_1^2 - \frac{(\Sigma X)^2}{n_1}}{n_1 - 1} = \frac{254}{8} = 31.75$$

$$s_2^2 = \frac{\Sigma X_2^2 - \frac{(\Sigma X_2)^2}{n_2}}{n_2 - 1} = \frac{277.5}{5} = 55.5$$

$$F = \frac{s_2^2}{s_1^2} = \frac{55.5}{31.75} = 1.748 \quad \text{is the calculated F.}$$

$\nu_1 = n_2 - 1$ since sample 2 has the larger sample variance which goes in the numerator.

The table F value is $F_{(5,8,.05)} = 3.6875$; hence

R: F > 3.6875

Since we are not in the rejection region, there is no evidence to show that the variances are not equal.

16. **Solution**

H_o: $\sigma_1^2 = \sigma_2^2$

H_a: $\sigma_1^2 \neq \sigma_2^2$

$\alpha = .05$, $\alpha/2 = .025$

$s_1^2 = 3.4^2 = 11.56$, $s_2^2 = 3.8^2 = 14.44$; hence

$$F = \frac{s_2^2}{s_1^2} = \frac{14.44}{11.56} = 1.249$$

Since s_2^2 is larger, $\nu_1 = n_2 - 1 = 15$, $\nu_2 = n_1 - 1 = 19$; we have $F_{(15,19,.025)} = 2.6171$.

The calculated F = 1.249 is less than the table F = 2.6171; therefore, do not reject H_o. There is no evidence to indicate that the variances are not equal.

Analysis of Variance

Major Topics and Key Concepts

10.1 Review of Hypothesis Testing

- Analysis of variance provides a method for testing the null hypothesis that three or more population means are equal.

10.2 The Analysis of Variance Computations

- Assumptions necessary in order to use this procedure.
- Between groups variance, s_B^2: measures how far the sample means deviate from the grand mean.
- Within groups variance, s_W^2: pooled estimate of the common variance for all the populations; measures the random variation within each of the samples.

10.3 Testing the Hypothesis of Several Equal Means

- Hypotheses: H_0: all population means are equal

  H_a: at least one mean is different.

- Logic: If all population means <u>are</u> equal, then s_B^2 should be close in value to s_W^2. If at least one population mean is different from the others, then expect s_B^2 to be larger than s_W^2.

- $F = s_B^2/s_W^2$ measures the ratio of the two variances.
- Use Table VIII to determine the rejection region.
- The Analysis of Variance Table: A summary table for presenting computations.

10.4 Procedures for Handling Unequal Sample Sizes

10.5 Another View of Analysis of Variance

- Total Sum of Squares

- Between Groups Sum of Squares and Between Groups Mean Square.
- Within Groups Sum of Squares and Within Groups Mean Square.

10.6 Analysis of Variance and Two-Sample Tests

- Equivalence of this method with that of Section 9.8.

10.7 Summary

- Table 10.3 contains all necessary formulas.
- Observe the following relationships:

Between groups SS = $\Sigma n_j (\bar{X}_j - \bar{\bar{X}})^2$

s_B^2 = Between groups MS = $\dfrac{\text{Between groups SS}}{k-1}$.

Within groups SS = $\Sigma (n_j - 1) s_j^2$ or $\sum_j \sum_i (X_{ij} - \bar{X}_j)^2$

s_W^2 = Within groups MS = $\dfrac{\text{Within groups SS}}{\Sigma n_j - k}$.

CHAPTER 10: Review Test

1. The technique of analysis of variance provides a method for extending the two sample hypothesis tests for _____ to more than two independent samples.
   (a) Population proportions
   (b) Population means
   (c) Population variances
   (d) Population sizes

2. In order to use analysis of variance, which of the following assumptions does not have to be satisfied?
   (a) Each of the k populations from which the samples were taken must be normally distributed.
   (b) All k populations must have the same population variances.
   (c) All k populations must have the same population means.
   (d) The k samples must be independent of each other.

3. The null hypothesis in an analysis of variance problem using k groups is always
   (a) $H_o: \mu_1 = \mu_2 = \ldots = \mu_k$ (i.e., all equal to each other).
   (b) $H_o: \mu_1 = \mu_2 = \ldots = \mu_k = \mu_o$ (i.e., all means are equal to some specified value).
   (c) $H_o: \mu_1 \neq \mu_2 \neq \ldots \neq \mu_k$ (i.e., none of the means are equal to each other).
   (d) None of the above.

4. If the null hypothesis is rejected then the conclusion must be
   (a) All of the population means are different.
   (b) At least one of the population means is different from the others.
   (c) All of the population means are equal.
   (d) None of the above.

5. The value of s_B^2, which is equivalent to the between group mean square, measures
   (a) How close the individual observations are to the grand mean.
   (b) How close the individual observations in each sample are to their sample mean.
   (c) How close together the k sample means are.
   (d) None of the above.

6. The value of s_W^2, which is equivalent to the within groups mean square, measures
   (a) How close the individual observations are to the grand mean.
   (b) How close the individual observations in each sample are to their sample mean.
   (c) How close together the k sample means are.
   (d) None of the above.

7. A "mean square" is equal to a "sum of squares of deviations from a mean" divided by the appropriate number of degrees of freedom.
   (a) True.
   (b) False.

8. If the true population means are not all equal, then we would expect that
   (a) s_B^2 would be equal to s_W^2.
   (b) s_B^2 would be less than s_W^2.
   (c) s_B^2 would be greater than s_W^2.
   (d) s_B^2 would be equal to F.

9. The F statistic of analysis of variance is always constructed by
   (a) Putting the larger of s_B^2 and s_W^2 in the numerator as was done in Chapter 9.
   (b) Dividing s_B^2 by s_W^2.
   (c) Dividing s_W^2 by s_B^2.
   (d) None of the above.

10. The null hypothesis should be rejected at the alpha level whenever the calculated F is
    (a) Greater than $F(\alpha/2, k-1, \Sigma n_i - k)$.
    (b) Greater than $F(\alpha, k-1, \Sigma n_i - k)$.
    (c) Greater than $F(\alpha, \Sigma n_i - k, k-1)$.
    (d) Greater than $F(\alpha/2, \Sigma n_i - k, k-1)$.

11. Total sum of squares is equal to
    (a) $s_B^2 + s_W^2$.
    (b) $\nu_1 + \nu_2$.
    (c) Between groups mean square + within groups mean square.
    (d) Between groups sum of squares + within groups sum of squares.

12. If five samples of sizes 5, 9, 4, 5 and 6 are taken from five normal populations, then rejection region for $\alpha = .01$ is
    (a) $F > 3.7254$.
    (b) $F > 13.929$.
    (c) $F > 3.8951$.
    (d) $F > 4.2184$.

13. A within groups mean square equal to zero would imply that
    (a) The sample sizes were all equal.
    (b) All the sample means were equal.
    (c) All the observations within each of the samples were all equal.
    (d) All the observations in all of the samples were equal.

CHAPTER 10: Answers to Review Test

Question	Answer	Text Section Reference
1	b	10.2
2	c	10.2
3	a	10.3
4	b	10.3
5	c	10.3
6	b	10.3
7	a	10.5
8	c	10.3
9	b	10.3
10	b	10.3
11	d	10.5
12	d	10.3
13	c	10.5

CHAPTER 10: Review Problems

1. An auto rental company has a large fleet consisting of four types of automobiles in one class. When originally purchased, all four types of cars in this class were advertised as having the same expected gasoline mileage. After the first year of ownership, a random sample of each type of car was taken and the gas mileage was measured. Determine if the original claim is still true using .05 level of significance.

Auto Type	Gasoline Mileages
1	20.0 21.2 18.4 20.5 21.4 20.3 19.5 20.3
2	16.4 20.8 22.4 21.0 20.2
3	24.0 20.0 23.5 24.2 23.9 23.8 23.0
4	15.8 22.4 20.8 22.6

2. The Environmental Protection Agency, EPA, has strict rules governing the amount of industrial waste that can be released into rivers. In order to comply with regulations a manufacturer experimented with five different methods of restricting the amount of chemicals that were released from the plant's cooling process. Nine measurements were made using each method. The data were recorded for a particular type of chemical waste still present in the water after treatment and is measured in parts per million (ppm). Determine if there are any differences in the true mean amount of chemicals in the water for the five methods. Use an alpha of .05.

Method	Sample Mean	Sample Variance
1	110	30
2	104	34
3	114	28
4	90	36
5	94	20

3. Complete the following analysis of variance table.

Source of Variation	df	SS	MS	F
Between groups	12			3
Within groups	__	__	12	
Total	36	720		

   (a) How many populations are being compared?

(b) Were the sample sizes all equal?
(c) A necessary assumption for using analysis of variance is that each population has the same variance. What is the best estimate of this common variance?
(d) What conclusion should be made at the .025 level?

4. Three different plans have been suggested for routing a product through an assembly line. The ideal plan would be the one requiring the least amount of time. A small scale experiment is conducted using each of the methods. Adopting plan C would be very expensive since major retooling would be required. Plans A and B would require only minor production changes. As a plant manager in charge of production, which plan would you adopt if your alpha level is .01? All measurements are in minutes.

Plan:	A	B	C
	$n_1 = 20$	$n_2 = 25$	$n_3 = 18$
	$\bar{X}_1 = 44.17$	$\bar{X}_2 = 43.00$	$\bar{X}_3 = 41.00$
	$s_1 = 3.20$	$s_2 = 2.82$	$s_3 = 3.60$

5. Consider four methods of "dialing" a telephone number: conventional dial, push tone, princess phones with small buttons and mobile phones. A random sample of 32 phone numbers were randomly selected from a city directory and eight were randomly assigned to each "dialing" mechanism. The time from start until the first ring was measured. Is any method faster than the others? Use the .05 level of significance.

Method	Times (in seconds)
Dial	16 14 18 16 14 14 13 13
Push tone	14 12 15 10 13 12 15 11
Princess	20 17 17 14 16 16 17 15
Mobile	22 20 18 16 19 18 20 17

6. The marketing research department for a chain of convenience stores tested the effects of five different floor displays for records and tapes. Thirty comparable stores were randomly selected for use in the study and each display was randomly assigned to six stores. The sales for the records and tapes for one week period were recorded. Unfortunately, three store managers forgot to set up the display and in another store the entire

display was stolen. Which display would you recommend for nation-wide use? Use alpha equal to .05.

	\multicolumn{5}{c}{Display}
	1	2	3	4	5
One week sales for records and tapes.	221	200	96	236	126
	210	240	218	343	X
	320	225	X	111	284
	120	260	324	X	X
	180	125	174	142	172
	254	180	118	138	222

CHAPTER 10: Solutions to Review Problems

1. **Solution**

   H_o: $\mu_1 = \mu_2 = \mu_3 = \mu_4$.
   H_a: at least one population mean is different.
   Use the formula best suited to the form in which the data is presented to you. Here we have raw data given.

   Summarize the Data

Auto Type	n_j	Sample Totals	$\bar{X}_j$	$n_j(\bar{X}_j - \bar{\bar{X}})^2$	$\sum_i (X_{ij} - \bar{X}_j)^2$
1	8	161.6	20.20	6.48	6.32
2	5	100.8	20.16	4.418	20.272
3	7	162.4	23.20	30.87	12.86
4	4	81.6	20.40	1.96	30.16
	24	506.4 (grand total)	$\bar{\bar{X}}$ = 21.10* (grand mean)	43.728 (between groups SS)	69.612 (within groups SS)

   *Note, the grand mean is <u>not</u> the average of the sample means unless the sample sizes are all equal.

   Analysis of Variance Table

Source	df	SS	MS	F
Between auto types	3	43.728	14.5760	4.188
Within auto types	20	69.612	3.4806	
Total	23	113.340		

   The table F is $F_{(.05, 3, 20)} = 3.0984$.

   Conclusion

   Since the calculated F = 4.188 exceeds the table F, we reject H_o.

   The true mean gasoline mileage for the four types of rental cars in this class are not all equal.

2. **Solution**

H_o: $\mu_1 = \mu_2 = \mu_3 = \mu_4 = \mu_5$.
H_a: at least one mean is different.
The data are given to in summarized form.

Calculations

Method	n_j	$\bar{X}_j$	$n_j(\bar{X}_j - \bar{\bar{X}})^2$	s_j^2	$(n_j-1)s_j^2$
1	9	110	519.84	30	240
2	9	104	23.04	34	272
3	9	114	1211.04	28	224
4	9	90	1383.84	36	288
5	9	94	635.04	30	240
	45	$\bar{\bar{X}} = 102.4$	3772.80		1264

↗ between groups SS ↗ within groups SS

Analysis of Variance Table

Source	df	SS	MS	F
Between methods	4	3772.8	943.2	29.848
Within methods	40	1264.0	31.6	
Total	44	5036.8		

Table F value is $F_{(.05, 4, 40)} = 2.6060$.

Conclusion

Since the calculated F = 29.848 is greater than the table F value, we reject H_o.

At least one method has a population mean which is different from the others.

3. **Solution**

Source of Variation	df	SS	MS	F
Between groups	12	432	36	3
Within groups	24	288	12	
Total	36	720		

a. Since k-1 = 12, then k = 13 populations.

b. No. There is a total of 37 observations for 13 populations; hence, the sample sizes cannot all be equal.

c. s_W^2 = within groups MS = 12 is the best estimate of σ^2.

d. Since $F_{(.025,12,24)}$ = 2.5412, and the calculated F = 3, we would reject H_o and claim that at least one population mean is different from the others.

4. **Solution**

H_o: $\mu_1 = \mu_2 = \mu_3$.
H_a: at least one mean is different.

For this problem, we will demonstrate the formulas directly. In order to calculate s_B^2 we need the grand mean. Since the sample sizes are not equal, this is a weighted average of the sample means.

$$\bar{\bar{X}} = \frac{\Sigma n_i \bar{X}_i}{\Sigma n_i} = (20 \cdot 44.17 + 25 \cdot 43.0 + 18 \cdot 41.0)/63$$

$$= 42.8$$

Now, $s_B^2 = \frac{\Sigma n_j (\bar{X}_j - \bar{\bar{X}})^2}{k-1}$

$$= [20(1.37)^2 + 25(0.2)^2 + 18(-1.8)^2]/2$$

$$= 96.858/2 = \boxed{48.429}$$

Note that we are given standard deviations and not variances for each group.

$s_W^2 = \frac{\Sigma (n_j - 1) s_j^2}{\Sigma n_j - 1}$

$$= [19(3.2)^2 + 24(2.82)^2 + 17(3.6)^2]/60$$

$$= 605.7376/60 = \boxed{10.0956}$$

Expressing these values in an analysis of variance of table,

Source	df	SS	MS	F
Between plans	2	96.8580	48.4290	4.797
Within plans	60	605.7376	10.0956	
Total	62	702.5956		

The table F value is $F_{(.01, 2, 60)} = 4.9774$.

Since the calculated $F = s_B^2/s_W^2 = 48.4290/10.0956 = 4.797$ is less than the table value, we do not reject H_0.

Hence, we conclude that there is no evidence to indicate that at least one of the population means is different from the others. The production manager could recommend any of the three plans, but probably A or B since they are the least costly to adopt.

One could argue that C has the lowest sample mean and should, therefore, be used. However, this was probably due to sampling variation because the differences were not significant; if the experiment were conducted again, it would not be unlikely for the order of the means reversed.

5. Solution

H_0: $\mu_1 = \mu_2 = \mu_3 = \mu_4$.
H_a: at least one is different.

Analysis of Variance Table

Source of Variation	df	SS	MS	F
Between methods	3	156.375	52.125	15.778
Within methods	28	92.500	3.304	
Total	31	248.875		

i.e., $s_B^2 = 52.125$, $s_W^2 = 3.304$,

$$F = s_B^2/s_W^2 = 15.778$$

$F_{(.05,3,28)} = 2.9467$; therefore, reject H_o.
The true mean times for the four dialing methods are not all equal.

6. Solution

   H_o: $\mu_1 = \mu_2 = \mu_3 = \mu_4 = \mu_5$.
   H_a: at least one mean is different.

   Ignore the stores for which no data are available and reduce the sample sizes accordingly.

   Analysis of Variance Table

Source of Variation	df	SS	MS	F
Between displays	4	3,093.0	773.250	.1377
Within displays	21	117,945.5	5616.452	
Total	25	121,038.5		

   $s_B^2 = 773.250$, $s_W^2 = 5616.452$, $F = s_B^2/s_W^2 = .1377$

   $F_{(.05,4,21)} = 2.8401$

   There is no evidence to indicate that any display is more effective than the others. It does not matter which display is used.

Approximate Tests: Multinomial Data

Major Topics and Key Concepts

- Three types of problems

    1. Testing two or more population proportions equal.
    2. Goodness of fit.
    3. Contingency table analysis.

11.1 The Multinomial Distribution

- An extension of the binomial to more than two possible classes (or responses) for each trial.
- k = number of possible classes that could occur.
- n = the number of trials.
- p_i = true probability of ith class occurring,
$$\sum_{i=1}^{k} p_i = 1.$$
- Y_i = <u>count</u> of the number of occurrences of the ith class in n trials, $\Sigma Y_i = n$.
- $f_i = Y_i/n$ = proportion of the n trials that fall into class i,
$$\sum_{i=1}^{k} f_i = 1.$$

11.2 A Hypothesis About the Multinomial Parameters

- H_o: $p_i = p_{io}$, for i = 1,...,k ⇒ the true population proportions equal some hypothesized values, p_{io}, for each of the k classes.
- $E_i = np_{io}$ ⇒ the number of observations out of n trials that we would <u>expect</u> to fall in class i, if H_o were really true.
- Y_i = the number of observations out of n trials that were <u>actually</u> observed in class i.

- Test statistic

$$\chi^2 = \sum_{i=1}^{k} \frac{(Y_i - E_i)^2}{E_i} \Rightarrow \text{a measure of how close } Y_i \text{ is to } E_i \text{ for each class.}$$

- The test statistic follows approximately a <u>Chi-Square distribution</u> with k-1 degrees of freedom.
- For <u>the approximation</u> to be good, $E_i \geq 5$ should be true for all classes.
- Use Table IX to determine the rejection region.
- Rejection region is always in the upper tail.

11.3 Binomial Data

- Relationship between sections 9.2 and 11.2.

11.4 A Test for Goodness of Fit

- A method for determining if a population follows a specific probability distribution such as a normal distribution.
- Possible need for combining classes to insure $E_i \geq 5$ for most classes.

11.5 Contingency Tables

- A method for determining if two categorical type random variables are dependent upon each other.
- Independence $\Rightarrow$ the proportion of each row total that belongs in the jth column is the same for all rows (and vice versa).
- Hypotheses: H_o: The two variables are independent.
  H_a: The two variables are dependent.
- Test statistics:

$$\chi^2 = \sum_{i,j} \frac{(Y_{ij} - E_{ij})^2}{E_{ij}} \quad \text{where } E_{ij} = R_i C_j / n.$$

- Use Table IX, with (r-1)(c-1) degrees of freedom for rejection region.

11.6 Summary

CHAPTER 11: Review Test

1. A multinomial population is one in which the elements are classified as belonging to one of k classes, where $k \geq 2$, and the appropriate model is the multinomial distribution if:

    (a) It is derived from a binomial distribution.
    (b) The probability of obtaining an element from a given class remains relatively constant.
    (c) The proportions of the elements belonging to each class are not changed by the selection of the sample.
    (d) It is a disjoint distribution for the random variables.

2. The parameters of the multinomial distribution are:

    (a) Sample size.
    (b) Proportion of elements that belong to each class.
    (c) The means.
    (d) All of the above.
    (e) (a) and (b) only.

3. The squared difference between the observed and expected numbers in a given class divided by the observed number is denoted by χ^2 (chi-square).

    (a) True.
    (b) False.

4. For the hypothesis $H_o: p_i = p_{io}$, $i = 1,\ldots,k$, chi-square is:

    (a) A measure of the lack of agreement between the data and the hypothesis.
    (b) A measure of the agreement between H_o and H_a.
    (c) Irrelevant.
    (d) An approximation of the agreement between the hypothesis and data.

5. For testing $H_o: p_i = p_{io}$, $i = 1,\ldots,k$, the degrees of freedom associated with a chi-square distribution are:

    (a) Dependent on the size of the sample.
    (b) Equal to one less than the number of classes.
    (c) Equal to the number of classes.
    (d) Equal to one plus the number of classes.

6. Which of the following statements is not true of the chi-square distribution?
    (a) It is a continuous distribution ordinarly derived as the sampling distribution of a sum of squares of independent standard normal variables.
    (b) It is a skewed distribution.
    (c) It is dependent upon two parameters, sample size and the degrees of freedom.
    (d) Its observed values are never negative.

7. Perfect agreement between the observed and expected numbers would result in a χ^2 value of
    (a) 0
    (b) 1
    (c) ∞
    (d) None of the above.

8. In hypothesis testing, the chi-square test is always a one-tailed test on the upper tail of the χ^2 distribution; the critical region is:
    (a) $\chi^2 \leq \chi^2_{\alpha, k-1}$
    (b) $\chi^2 \geq \chi^2_{\alpha/2, k}$
    (c) $\chi^2 \geq \chi^2_{\alpha, k-1}$
    (d) $\chi^2 \leq \chi^2_{\alpha/2, k-1}$

9. The multinomial chi-square may be used to test the goodness of fit of data with
    (a) Normal distributions.
    (b) Binomial distributions.
    (c) Poisson distributions.
    (d) All of the above.

10. In using the test for goodness of fit, each expected number should be at least _____, so that the chi-square statistic closely follows the chi-square distribution.
    (a) 10   (c) 3
    (b) 5   (d) 1

11. In tests for goodness of fit, one degree of freedom is lost for every parameter of the original population that must be estimated from the sample.
    (a) True.
    (b) False.

12. When testing the null hypothesis that the original population follows a normal distribution, the correct number of degrees of freedom for the goodness of fit test will be

   (a) k - 1 if both μ and σ² are specified in H_o.
   (b) k - 2 if σ² is specified in H_o but μ is estimated with the sample data.
   (c) k - 3 if neither μ nor σ² are specified in H_o and both must be estimated with the sample data.
   (d) All of the above are correct.

13. The null hypothesis that two categorical variables are independent of each other can be tested by using

   (a) The chi-square test.
   (b) A goodness of fit test.
   (c) Contingency tables.
   (d) Point estimation.

14. Independence in a contingency table when the proportion that each row total makes up is equal to the proportion that each corresponding column total makes up.

   (a) True.
   (b) False.

15. The calculated χ^2 in tests for independence in contingency tables has _____ degrees of freedom.

   (a) n - 1
   (b) (r - 1)(c - 1)
   (c) k - 1
   (d) k - 2

16. An 8x6 contingency table has _____ degrees of freedom.

   (a) 14   (c) 35
   (b) 12   (d) 48

17. The simplest contingency table, the 2 by 2, can be used to test the hypothesis that two binomial populations have the same relative frequency of success.

   (a) True.
   (b) False.

CHAPTER 11: Answers to Review Test

Question	Answer	Text Section Reference
1	c	11.1
2	e	11.1
3	b	11.2
4	a	11.2
5	b	11.2
6	c	11.2
7	a	11.2
8	c	11.2
9	d	11.4
10	b	11.4
11	a	11.4
12	d	11.4
13	c	11.5
14	b	11.5
15	b	11.5
16	c	11.5
17	a	11.5

CHAPTER 11: Review Problems

1. The safety department records of a western state show that the last 315 traffic fatalities occurred on the following week days:

Monday	35	Friday	55
Tuesday	25	Saturday	70
Wednesday	30	Sunday	60
Thursday	40		

   Test the hypothesis that traffic fatalities are uniformly distributed throughout the week. Let $\alpha = .01$.

2. The distribution of television viewers during the 8:00 p.m. to 9:00 p.m. CST slot on Mondays in previous weeks is given below. A random sample of 900 viewers this week resulted in the distribution on the right. Using the .01 level of significance, determine if this week's distribution fits the pattern of previous weeks.

	Previous Weeks	This Week
ABC	30%	28%
CBS	28%	25%
NBC	24%	28%
Public TV	4%	6%
Other	14%	13%

3. In testing a hypothesis about the mean length of office phone calls it is necessary to assume that the population of all phone call times follows a normal distribution. Using the sample data below, determine if this assumption is reasonable using $\alpha = .05$.

Length of Call (in min.)	Number of calls
0 but less than 2	6
2 but less than 4	12
4 but less than 6	33
6 but less than 8	43
8 but less than 12	32
12 but less than 16	15
16 but less than 20	9
	150

4. A plant manager believes that the number of industrial accidents per week follows a poisson distribution. Using $\alpha = .05$, test his claim.

No. of accidents	0	1	2	3	4	5	6
Observed Number	10	20	40	18	8	0	4

5. Samples of three kinds of materials that were subjected to extreme temperature changes produced the following results:

	Material 1	Material 2	Material 3
Broke Completely	25	45	41
Showed Slight Defects	40	35	33
Remained Perfect	35	20	26

Determine if there exists a dependence between the types of materials and the effects of extreme temperatures. Use the .05 level of significance.

6. In a nationwide survey, 200 respondents are asked if they feel that the current national energy program is effective. Determine if there is a dependence between the response given and the education level of the respondents, using the .01 level of significance.

Education	ENERGY PROGRAM Is Effective	Is Not Effective	
1. High school graduate or less	8	8	16
2. H.S. with at least 1 year of college	50	34	84
3. College graduate	17	58	75
4. Higher degree	5	20	25
	80	120	200

CHAPTER 11: Solutions to Review Problems

1. **Solution**

   If the fatalities are uniformly distributed over each of the seven days, then we would expect 1/7 of the fatalities to occur on any given day.

   H_o: $p_1 = p_2 = p_3 = p_4 = p_5 = p_6 = p_7 = 1/7$.
   H_a: at least one of the $p_i \neq 1/7$.

   Since $n = 315$ and $p_i = 1/7$ if H_o is true, then the expected number of fatalities is $E_i = n \cdot p_i = 45$ for each day of the week. Therefore,

   $$\chi^2 = \sum_{i=1}^{7} \frac{(Y_i - E_i)^2}{E_i} = \frac{(35-45)^2}{45} + \frac{(25-45)^2}{45}$$
   $$+ \frac{(30-45)^2}{45} + \frac{(40-45)^2}{45} + \frac{(55-45)^2}{45} + \frac{(70-45)^2}{45}$$
   $$+ \frac{(60-45)^2}{45} = \frac{1700}{45} = 37.78.$$

   From Table IX, $\chi^2_{.01, 6} = 16.81$.

   The calculated χ^2 is greater than the table value, therefore we reject the null hypothesis and conclude that the fatalities are not uniformly distributed. Weekends seem to have more accidents.

2. **Solution**

   H_o: The distribution this week "fits" the same distribution of previous weeks; i.e., there is no change.
   H_a: This week's distribution does _not_ fit previous weeks.

   Note: _Never_ calculate the χ^2 statistic using _percentages_; always convert to raw data, or counts. Since $n = 900$ and each p_{io} is given, we calculate $E_i = n \cdot p_{io}$ and $Y_i = n \cdot f_i$.

	Previous Weeks E_i	This Week Y_i	$Y_i - E_i$	$\dfrac{(Y_i - E_i)^2}{E_i}$
ABC	270	252	-18	1.2000
CBS	252	225	-27	2.8929
NBC	216	252	36	6.0000
Public TV	36	54	18	9.0000
Other	126	117	- 9	.6429
	900	900	0	19.7358 = χ^2

From Table IX, $\chi^2_{(.01,4)} = 13.28$

Since the calculated $\chi^2 = 19.7358$ is greater than the table value, we reject H_o and conclude that the percentages watching the different networks has changed.

3. Solution

H_o: Population of times <u>is</u> normally distributed.

H_a: Population of times <u>is not</u> normally distributed.

In the previous two problems the p_{i_0} values were easily determined (since they were given to us). Now, we must use the normal tables in order to calculate the hypothesized probabilities for each class. (Recall Chapter 5.) Before we can transform to the standard normal distribution we need to estimate μ and σ. Since the data are grouped, we calculate $\bar{X}$ and S as we did in Chapter 3.

Class Midpoint, V_i	Frequencies, f_i	$V_i f_i$	V_i^2	$f_i V_i^2$
1	6	6	1	6
3	12	36	9	108
5	33	165	25	825
7	43	301	49	2107
10	32	320	100	3200
14	15	210	196	2940
18	9	162	324	2916
	150	1200		12102

① $\bar{X} = \dfrac{\Sigma f_i V_i}{n} = \dfrac{1200}{150} = 8.00$

$$s^2 = \frac{\Sigma f_i V_i^2 - \frac{(\Sigma f_i V_i)^2}{n}}{n-1} = \frac{12102 - \frac{(1200)^2}{150}}{149} = 16.7919$$

$$s = \sqrt{16.7919} = 4.098$$

We now transform the class boundaries to Z-values using $Z = (X - \bar{X})/s$, and then use Table V to determine the probability of each class occurring.

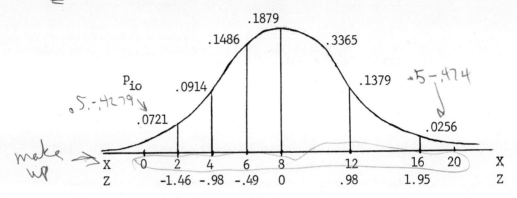

Since the sum of the probability must equal 1.0, do not use the highest and lowest class boundaries.

Finally, we have the hypothesized p_{io} values. Now,

Class	p_{io}	$E_i = n \cdot p_{io}$	$Y_i = f_i$	$Y_i - E_i$	$(Y_i - E_i)^2 / E_i$
less than 2.0	.0721	10.815	6	- 4.815	2.1437
2 to 4	.0914	13.710	12	- 1.710	.2133
4 to 6	.1486	22.290	33	10.710	5.1460
6 to 8	.1879	28.185	43	14.815	7.7873
8 to 12	.3365	50.475	32	-18.475	6.7623
12 to 16	.1379	20.685	15	- 5.685	1.5624
16 or more	.0256	3.840*	9	5.160	6.9227
	1.0000	150.000	150.	0	30.5487 = χ^2

*For this class E_i is less than 5. Since this was the only E_i value less than 5 and since it was just a little less than 5, we ignored the guideline. The alternative approach would be to combine the last two classes ($E_6 = 24.525$ and $Y_6 = 24$) and recompute the χ^2 value. Note that this would reduce the degree of freedom to 3.

The degrees of freedom are 7-1-2 = 4 since two parameters, μ and σ, had to be estimated; hence, $\chi^2_{(.05,4)} = 9.49$.

Since the calculated $\chi^2 = 30.5487$ is greater than 9.49, we reject H_o and conclude that the true population of lengths of office phone calls does not follow a normal distribution.

4. **Solution**

   H_o: Population fits a Poisson distribution.
   H_a: Population does not fit a Poisson distribution.

   We must estimate the mean of the Poisson distribution before we can find the p_{io} values in Table IV.

   $$\bar{X} = \frac{\Sigma f_i V_i}{\Sigma f_i}$$

   $= (10(0) + 20(1) + 40(2) + 18(3) + 8(4)$
   $\quad 0(5) + 4(6))/100.$

   $= 2.1$

   We can now enter Table IV using $\theta = 2.1$ to obtain p_{io}.

No. of Accidents	P_{io}	$E_i = n \cdot P_{io}$	Y_i	$(Y_i - E_i)^2 / E_i$
0	.1225	12.25	10	.4133
1	.2572	25.72	20	1.2721
2	.2700	27.00	40	6.2593
3	.1890	18.90	18	.0429
4	.0992	9.92	8	.3716
5	.0417	[4.17] * → 6.21	[0] → 4	.7865
6 or more	.0204	[2.04]	[4]	
		100	100	9.1457 = χ^2

*Since both E_i are less than 5 we combine these two classes. The degrees of freedom are 6-1-1 = 4 (since there are now only 6 classes and θ has to be estimated from the sample data). Now, $\chi^2_{(.05,4)} = 9.49$.

Since the calculated $\chi^2 = 9.1457$ is less than the table value we do not reject H_0. We conclude that there is no evidence to show that the distribution is not Poisson.

5. Solution

H_0: The type of material is <u>independent</u> of the effects of extreme temperatures.

H_a: The type of material is <u>dependent</u> on the effects of extreme temperatures.

$E_{ij} = r_i c_j / n$; i.e., $E_{11} = 111 \cdot 100/300 = 37$,

$E_{32} = \dfrac{81 \cdot 100}{300} = 27$, etc.

	Observed				Expected (if H_0 is true)			
Material	1	2	3		1	2	2	
Broke	25	45	41	111	37	37	37	111
Defects	40	35	33	108	36	36	36	108
Perfect	35	20	26	81	27	27	27	81
	100	100	100	300	100	100	100	300

$$\chi^2 = \dfrac{(25-37)^2}{37} + \dfrac{(45-37)^2}{37} + \dfrac{(41-37)^2}{37} + \dfrac{(40-36)^2}{36}$$

$$+ \dfrac{(35-36)^2}{36} + \dfrac{(33-36)^2}{36} + \dfrac{(35-27)^2}{27} + \dfrac{(20-27)^2}{27}$$

$$+ \dfrac{(26-27)^2}{27} = 10.998.$$

The degree of freedom are $(3-1) \cdot (3-1) = 4$; hence, $\chi^2_{(.05, 4)} = 9.49$. Since the computed $\chi^2 = 10.998$ is greater than the table value, we reject H_0. Our conclusion is that there is a dependence between the type of material and the effects at extreme temperatures; i.e., the materials do not behave the same when subjected to extreme temperatures.

6. Solution

H_0: Opinion is independent of education level.
H_a: Opinion is dependent on education level.

OBSERVED

Education	Program Is Effective	Is Not	
1	8	8	16
2	50	34	84
3	17	58	75
4	5	20	25
	80	120	200

EXPECTED (if H_o is true)

Education	Program Is Effective	Is Not	
1	6.4	9.6	16
2	33.6	50.4	84
3	30.0	45.0	75
4	10.0	15.0	25
	80	120	200

$$\chi^2 = \frac{(8-6.4)^2}{6.4} + \frac{(8-9.6)^2}{9.6} + \frac{(50-33.6)^2}{33.6} + \ldots$$

$$+ \frac{(20-15)^2}{15} = 27.56$$

The table χ^2 value is $\chi^2_{(.01,3)} = 11.34$.

Since the calculated $\chi^2 = 27.56$ exceeds the table value, we reject H_o. The option is influenced by the education level of the respondent.

Regression and Correlation

Major Topics and Key Concepts

- Regression analysis: determining the functional relationship between a dependent variable and one or more independent variables.

- Correlation analysis: measuring the strength of the relationship between two variables.

12.1 Simple Linear Regression

- "Simple" means only one independent variable.
- For each value of X, the independent variable, there is an entire population of values for Y.
- The means of all these populations are assumed to form a straight line.
- $\mu_{Y|X} = A + BX$ is the equation for the "true" regression line.
- A is the true intercept, B is the true slope; A and B are "parameters" and are estimated from the data.

12.2 Finding the Slope and Intercept of a Regression Line

- $\hat{Y} = a + bX$ is the estimator for $\mu_{Y|X}$.
- "a" is the sample intercept and estimates A; "b" is the sample slope and estimates B.
- <u>Least squares method</u> is used to find a and b; see equations 12.7 and 12.8.

12.3 Measuring the Accuracy of Prediction

- $\sigma^2_{Y|X}$ is the "true" variance of population of all Y values that can occur for a specific value of X.
- A small value of $\sigma^2_{Y|X}$ implies that the data points fall close to the true regression line.
- $s^2_{Y|X}$ is the estimator for $\sigma^2_{Y|X}$.
- A small value of $s^2_{Y|X}$ implies accurate predictions of Y when using the sample regression equation.
- Coefficient of determination, r^2: the proportion of the variation of the dependent variable, Y, that is explained using this X variable.

12.4 Statistical Inference in Regression Analysis

- Three assumptions.
- Confidence intervals for B and A; equations 12.13 and 12.17.
- Testing the hypotheses, H_o: $B = B_o$ and H_o: $A = A_o$; equations 12.15 and 12.19.
- Confidence intervals for the true mean of all Y values for a specified value of X, say X_o; equation 12.20.

12.5 Correlation

- ρ denotes the true population correlation between two variables.
- r denotes estimator of ρ.
- Testing the hypothesis, H_o: $\rho = 0$; equation 12.29.

12.6 Partitioning the Sum of Squares

12.7 The Relationship between Regression, Correlation, and Analysis of Variance

- Use of the analysis of variance table.

12.8 Regression, Correlation, and Computers

12.9 Summary

CHAPTER 12: Review Test

1. To use a regression analysis we must _____ the functional form of the relationship between the variables.

    (a) Be able to estimate
    (b) Know
    (c) Assume
    (d) Know or assume

2. If a functional relationship between variables exists, this implies that there is a cause-and-effect relationship.

    (a) True.
    (b) False.

3. The functional form of the relationship between variables to be used in regression analysis may be determined from:

    (a) Analytical or theoretical considerations.
    (b) Specially designed tables.
    (c) Scatter diagrams.
    (d) All of the above.
    (e) (a) and (c).

4. The regression curve or line is the curve or line that joins the _____ of the distributions corresponding to all possible values of X.

    (a) Means
    (b) Deviations
    (c) Points
    (d) Variances

5. In the regression prediction equation, $\hat{Y} = a + bX$, a and b are:

    (a) The intercept and slope, respectively.
    (b) Estimators for A and B, respectively, and together are used to estimate $\mu_{Y|X}$.
    (c) Obtained by use of least squares method.
    (d) All of the above.

6. The method of least squares will not give the best unbiased estimators for A and B if:

    (a) The X-values are nonrandom.
    (b) The X-values are random.
    (c) $\sigma^2_{Y \cdot X} = \sigma^2$ for all x.
    (d) For each value of X, Y is normally and independently distributed.

7. The least squares theory requires the assumption that each population of Y is normally distributed.

    (a) True.
    (b) False.

8. The regression equation represents a line that is fitted to the observed points so that the sum of squares of the vertical deviations from the line is:

    (a) Smaller than any other line could produce.
    (b) The greatest possible.
    (c) Equal to the cross product of X and Y.
    (d) None of the above.

9. In the regression equation, $\hat{Y} = 3.25 + 2.5X$, the B = 2.5 value indicates:

    (a) The value at which the line intercepts the X axis.
    (b) The starting point for the regression line.
    (c) That a one-unit change in X is associated with a 2.5 unit change in Y.
    (d) The value at which the line intercepts the Y axis.

10. The ratio of the sum of squares associated with the regression to the total sum of squares for Y is called _____, or r^2, and must always be _____.

    (a) The regression coefficient; greater than 0.
    (b) The correlation coefficient; less than 1.
    (c) The coefficient of determination; between 0 and 1.
    (d) The coefficient of alienation; between -1 and +1.

11. As the scatter of points about the regression line becomes greater, r^2 will:

    (a) Be unaffected.
    (b) Become smaller.
    (c) Become larger.
    (d) None of the above.

12. If r^2 is equal to 1.0, then we know that

    (a) a = 0.
    (b) b = 1.
    (c) $S^2_{Y|X} = 0$.
    (d) All of the above.

13. Hypothesis tests and confidence intervals for A and B require that the populations of Y values have a normal distribution.

    (a) True.
    (b) False.

14. The parameter of the joint distribution of two random variables X and Y that measures the degree of linear association between them is called:

(a) The regression coefficient.
(b) The coefficient of alienation.
(c) The coefficient of determination.
(d) The correlation coefficient.

15. A perfect linear relationship is indicated when ρ takes a value of:

    (a) 0
    (b) -1
    (c) +1
    (d) Both (a) and (c) could be correct.
    (e) Either (b) or (c) could be correct.

16. If one variable increases as the other decreases, then the variables are:

    (a) Positively correlated.
    (b) Negatively correlated.
    (c) Uncorrelated.
    (d) Not linearly associated.

17. If b = -1.4 and the coefficient of determination is .64, then the sample correlation will be

    (a) .64   (c) -.80
    (b) .80   (d) Cannot be determined.

18. In partitioning the sum of squares for a simple linear regression, the portion of the variation that is due to regression has _____ of freedom associated with it.

    (a) n-1 degrees.   (c) 1 degree.
    (b) n-2 degrees.   (d) n+1 degrees.

19. The unexplained, or random, variation is the variation of the observed Y's about the regression line. Its sum of squares has _____ of freedom.

    (a) n-2 degrees.
    (b) n-1 degrees.
    (c) n+1 degrees.
    (d) 1 degree.

20. The F-test for an analysis of variance is equivalent to the t-test for testing:

    (a) $H_o: B = B_o$ (any value).
    (b) $H_o: B = 0$.
    (c) $H_o: A = B = 0$.
    (d) None of the above.

CHAPTER 12: Answers to Review Test

Question	Answer	Text Section Reference
1	d	12.1
2	b	12.1
3	e	12.1
4	a	12.1
5	d	12.2
6	b	12.2
7	b	12.2
8	a	12.2
9	c	12.2
10	c	12.3
11	b	12.3
12	c	12.3
13	a	12.4
14	d	12.5
15	e	12.5
16	b	12.5
17	c	12.5
18	c	12.6
19	a	12.6
20	b	12.7

CHAPTER 12: Review Problems

1. From a random sample of 10 movie theaters, the box office gross income (in thousands of dollars) and the revenue from concessions (in hundreds of dollars) were measured for one week.

Theater	1	2	3	4	5	6	7	8	9	10
Box Office (X)	23	7	15	17	23	22	10	14	20	19
Concessions (Y)	11	3	5	7	9	8	5	7	8	7

   (a) Calculate the prediction equation.
   (b) What proportion of the total variation in concessions can be explained by using box office gross?
   (c) Estimate the standard deviation of Y around the regression line.
   (d) Predict the concession sales for a week when the box office gross is $21,500.
   (e) Construct a 90% confidence interval for the mean concession sales when box office gross is $21,500.

2. The relationship between the freshness of peaches and the time from picking was studied by randomly selecting twelve peaches from the same orchard. Every two days, two peaches were randomly selected and their freshness was measured on a 10-point scale. A score of 10 is perfect, 0 is rotten.

Day	Freshness
0	9.4 , 9.8
2	8.9 , 8.2
4	7.2 , 7.0
6	5.8 , 6.8
8	5.0 , 4.6
10	4.1 , 3.5

   (a) Determine the equation for predicting freshness.
   (b) Interpret the meaning of your value of b.
   (c) Construct a 95% confidence interval for the true slope.
   (d) Predict the freshness of a peach 18 days after picking. Explain what this value means.
   (e) Are these results valid for all peach orchards?

3. A retail grocer believes there is a linear relationship between the price per pound of bananas, X, and the weekly demand, Y. He collects data and summarizes as follows:

$\Sigma Y = 720$ $\Sigma X^2 - (\Sigma X)^2/n = 96$

$\Sigma Y = 240$ $\Sigma Y^2 - (\Sigma Y)^2/n = 30.5$

$n = 16$ weeks $\Sigma XY - (\Sigma X)(\Sigma Y)/n = -23.45$

Is there evidence to conclude that the demand decreases as price increases, using a .05 level of significance?

4. Suppose the raw data for a problem has been summarized such that $n = 10$, $\Sigma X^2 - (\Sigma X)^2/n = 25$, $\Sigma XY - (\Sigma X)(\Sigma Y) = 75$, and $s_{Y|X} = 2$. Using alpha of .10, test the hypothesis:

$$H_o: \beta = 4.0$$
$$H_a: \beta \neq 4.0 .$$

5. The data for eight grocery stores represent the number of jars of peanut butter sold in one day, X, and the number of loaves of bread sold during the same day, Y.

$\Sigma X = 168$, $\Sigma X^2 = 4518$, $\Sigma XY = 13066$

$\Sigma Y = 544$, $\Sigma Y^2 = 42530$, $n = 8$

(a) Calculate the sample correlation.
(b) Test the hypothesis that the true population correlation is zero. Use the .05 level of significance.

CHAPTER 12: Solutions to Review Problems

1. <u>Solution</u>

   Be sure that you see the distinction between "little" x_i and "capital" X_i in equations 12.7 and 12.9, and observe the relationship between them just before text example 2. For this problem, we will demonstrate the methods of equation 12.9; in all other problems we will use the second method.

Raw Data		Corrected for Mean				
X	Y	$x = X-\bar{X}$	$y = Y-\bar{Y}$	x^2	y^2	xy
23	11	6	4	36	16	24
7	3	-10	-4	100	16	40
15	5	-2	-2	4	4	4
⋮	⋮	⋮	⋮	⋮	⋮	⋮
19	7	2	0	4	0	0
$\Sigma X = 170$	$\Sigma Y = 70$	0	0	272	46	102
$\bar{X} = 17$	$\bar{Y} = 7$			Σx^2	Σy^2	Σxy

   Now that the data are reduced to the form in the box, we can proceed.

   a. $b = \dfrac{\Sigma xy}{\Sigma x^2} = \dfrac{102}{272} = .375$

   $a = \bar{Y} - b\bar{X} = 7 - .375(17) = .625$

   Therefore, the prediction equation is

   $$\hat{Y} = .625 + .375X \ .$$

   b. $r^2 = \dfrac{b\Sigma xy}{\Sigma y^2} = \dfrac{.375 \cdot 102}{46} = \dfrac{38.25}{46} = .8315 \ .$

   c. $s^2_{Y|X} = \dfrac{1}{n-2}[\Sigma y^2 - b\Sigma xy] = 1/8[46 - 38.25] = .96875 \ .$

   Hence, $s_{Y|X} = .98425$.

   d. When box office gross is \$21,500, then X = 21.5. Why? Therefore,

   $$\hat{Y} = .625 + .375 \cdot 21.5$$
   $$= 8.6875$$

   Hence, the predicted mean concession sales when weekly box office gross is \$21,500 is \$868.75.

e. This is from section 12.4.3. From part d,
$\hat{Y}_0 = 8.6875$; we need $s_{\hat{Y}_0}$ from equation 12.21.

$$s_{\hat{Y}_0} = s_{Y|X}\sqrt{\frac{1}{n} + \frac{(X_0 - \bar{X})^2}{\Sigma x^2}} = .98425\sqrt{\frac{1}{10} + \frac{(21.5 - 17)^2}{272}}$$

$$= .4111 .$$

For 90% confidence, $t_{(.05,8)} = 1.86$; the confidence interval is

$$\hat{Y}_0 \pm t_{(.05,8)} \cdot s_{\hat{Y}_0} = 8.6875 \pm 1.86 \cdot .4111$$

$$= 8.6875 \pm .7646$$

or (7.9229, 9.4521).

We are 90% confident that the true mean concession sales when box office gross is $21,500 will fall between $792.29 to $945.21.

2. **Solution**

Initially we must summarize data. Observe that day is X and freshness is Y since we wish to predict freshness for a given number of days.

X	Y	X²	Y²	XY
0	9.4	0	88.36	0
0	9.8	0	96.04	0
2	8.9	4	79.21	17.8
2	8.2	4	67.24	16.4
⋮	⋮	⋮	⋮	⋮
10	4.1	100	16.81	41.0
10	3.5	100	12.25	35.0

$\Sigma X = 60 \quad \Sigma Y = 80.3 \quad \Sigma X^2 = 440 \quad \Sigma Y^2 = 586.79 \quad \Sigma XY = 319.4$
$\bar{X} = 5.0 \quad \bar{Y} = 6.6917$

Hence,

$$\Sigma x^2 = \Sigma X^2 - (\Sigma X)^2/n = 140$$

$$\Sigma y^2 = \Sigma Y^2 - (\Sigma Y)^2/n = 49.449$$

$$\Sigma xy = \Sigma XY - (\Sigma X)(\Sigma Y)/n = -82.1 .$$

The data are now summarized; we can now proceed.

a. $b = \dfrac{\Sigma xy}{\Sigma x^2} = \dfrac{-82.1}{140} = -.5864$

$a = \bar{Y} - b\bar{X} = 9.6238$.

Therefore, $\hat{Y} = 9.6238 - .5864X$.

b. Each additional day the freshness <u>decreases</u> .5864 units.

c. For 95% confidence, $t_{(.025,10)} = 2.228$. We need s_b, equation 12.14.

$$s_b = \dfrac{s_{Y|X}}{\sqrt{\Sigma x^2}} \text{ where } s^2_{Y|X} = \dfrac{1}{10}[\Sigma y^2 - b\Sigma xy] = .13032$$

$$= \dfrac{.361}{\sqrt{140}} = .0305 .$$

Now the 95% confidence interval is

$$b \pm t_{(.025,10)} \cdot s_b$$

$$-.5864 \pm 2.228 \cdot .0305$$

$$(-.6544, -.5184)$$

I am 95% confident that the true slope falls in this interval.

d. For $X = 18$, $\hat{Y} = 9.6238 - .5864 \cdot 18 = -.93$.

The problem is that a predict equation is valid only within the range in which we have data points. We do not know what happens to the equation beyond 10; hence, we cannot predict freshness at 18 days with this equation.

e. Since the sample was taken only from only one orchard, the equation is valid only for peaches from that orchard.

3. Solution

Test $H_o: B \geq 0$

$H_a: B < 0$

$t_{(.05,14)} = 1.761$; hence,

R: $t < -1.761$.

From the data, we have

$$\Sigma x^2 = 96, \quad \Sigma y^2 = 30.5, \quad \Sigma xy = -23.45.$$

Now,

$$b = \frac{\Sigma xy}{\Sigma x^2} = -.24427$$

$$s^2_{Y|X} = \frac{1}{14}[\Sigma y^2 - b\Sigma xy] = 1.7694$$

$$s_{Y|X} = 1.330$$

$$s_b = \frac{s_{Y|X}}{\sqrt{\Sigma x^2}} = .1358.$$

Therefore, $t = \dfrac{b - B_o}{s_b} = -1.799$.

Since the calculated t is less than -1.761 (from table), we reject H_o. We conclude that demand <u>does</u> decrease as price increases.

4. <u>Solution</u>

$H_o: \beta = 4$

$H_a: \beta \neq 4$

$t_{(.05,8)} = 1.86$; hence,

R: $t < -1.86$ or $t > 1.86$.

From the data, $b = \dfrac{\Sigma xy}{\Sigma x^2} = \dfrac{75}{25} = 3.0$

$$s_b = \frac{s_{Y|X}}{\sqrt{\Sigma x^2}} = \frac{2}{5} = .4.$$

Therefore, $t = \dfrac{b - B_o}{s_b} = \dfrac{3-4}{.4} = -2.5$; we reject H_o.

The true slope is not equal to 4.0.

5. Solution

$$\Sigma x^2 = \Sigma X^2 - (\Sigma X)^2/n = 4518 - \frac{(168)^2}{8} = 990$$

$$\Sigma y^2 = \Sigma Y^2 - (\Sigma Y)^2/n = 42530 - \frac{(544)^2}{8} = 5538$$

$$\Sigma xy = \Sigma YX - (\Sigma X)(\Sigma Y)/n = 13066 - \frac{168 \cdot 544}{8} = 1642$$

a. $r = \dfrac{\Sigma xy}{\sqrt{\Sigma x^2 \cdot \Sigma y^2}} = \dfrac{1642}{\sqrt{990 \cdot 5538}} = .701$.

b. $H_o: \rho = 0$

$H_a: \rho \neq 0$

$t_{(.025, 6)} = 2.447$

The calculated t-value is

$t = r\sqrt{\dfrac{n-2}{1-r^2}} = 2.409$.

Since the calculated t-value is not in the rejection region, we do not reject H_o.

We conclude that there is insufficient evidence to indicate that the population correlation is anything other than zero.

You might argue that r = .701 seems very large; however, with the sample size, n = 8, the sampling variation is quite large.

13

Multiple Regression

Major Topics and Key Concepts

- Multiple means two or more independent variables are used to predict Y.
- $\mu_{Y|X_1\ldots X_k} = A + B_1X_1 + B_2X_2 + \ldots + B_kX_k$ represents the plane containing the true mean of the population of Y values for a set of values, $X_1, \ldots, X_k$.

13.1 Finding the Coefficients for a Multiple Regression Equation

- Normal equations.
- Need for computer.

13.2 Measuring the Accuracy of Predictions in Multiple Regression

- $s^2_{Y|1\ldots k}$ is the estimator for the true variance of the population of Y values that can occur for a specific set of values of $X_1, \ldots X_k$.
- Coefficient of multiple determination, R^2: the proportion of the variation of Y that is explained by using a set of X variables.

13.3 Statistical Inference in Multiple Regression

- Three assumptions.
- A confidence interval for an individual coefficient, B_i; equation 13.10.
- Testing the hypothesis, H_0: $B_i = 0$; equation 13.13.
- Multicollinearity: linear dependences that may exist among two or more of the independent variables.

13.4 The Beta Coefficients

14.5 Partitioning the Sum of Squares in Multiple Regression

- Use of the analysis of variance table.

- The overall F-test, H_o: $B_1 = \ldots = B_k = 0$.

13.6 Stepwise Multiple Regression

- First use the X variable that, by itself, does the best job of explaining the variation in Y.
- Add variables one at a time in order of their ability to explain the variation in Y.
- Stop when the next variable to be added does not sufficiently increase R^2.

13.7 A Computer Example of Stepwise Regression

13.8 Special Problems in Multiple Regression

- Violating assumptions.
- Over specification.
- Multicollinearity.
- Extrapolation.
- Cause-and-effect.

CHAPTER 13: Review Test

1. The word "multiple" in multiple regression means:
    (a) All the independent variables are multiplied together.
    (b) Each independent variable is multiplied by the dependent variable.
    (c) There are at least two independent variables used to predict the dependent variables.
    (d) None of the above.

2. The coefficients of the independent variables in a multiple regression are always exactly the same as they would be if several simple linear regressions were performed.
    (a) True.
    (b) False.

3. If the dependent variable, Y, is measured in dollars and X_2 is measured in barrels, then the estimated coefficient of X_2, denoted by b_2, is measured in:
    (a) Barrels.
    (b) Dollars.
    (c) Barrels per dollar.
    (d) Dollars per barrel.

4. The coefficient of multiple determination is measured in what units?
    (a) The same units as the dependent variable.
    (b) The same units as the last independent variable.
    (c) The square of the units of the dependent variable.
    (d) None of the above.

5. The proportion of squared error in estimating the Y_i values that can be eliminated by using this regression equation, i.e., the proportion of variation of Y_i values explained by this set of independent variables, is denoted by:
    (a) $s^2_{Y|12\cdots k}$.
    (b) $s_{Y|12\cdots k}$.
    (c) R^2.
    (d) R.

6. Which of the following is not a basic assumption needed for hypothesis testing for a multiple regression problem?
   (a) The independent variables are a nonrandom.
   (b) The dependent variable's values are normally and independently distributed around the regression plane.
   (c) The variance of the dependent variable around the regression plane is constant.
   (d) The true coefficients, B_i, are normally distributed.

7. If two independent variables are used to predict Y and if b_1 is three times larger than b_2, then X_1 is a more sensitive variable for explaining the variation of Y.
   (a) True.
   (b) False.

8. The number of standard deviations that $\hat{Y}$ changes for a one standard deviation change in X_4 is denoted by:
   (a) A confidence interval on B_4.
   (b) The beta coefficient, β_4.
   (c) The estimated coefficient, b_4.
   (d) The standard error of b_4, s_{b_4}.

9. The value of $s^2_{Y|12\cdots k}$ is equal to:
   (a) Mean square due to regression.
   (b) Mean square of deviations from regression, or mean square residual.
   (c) The overall F value.
   (d) The coefficient of multiple determination.

10. The overall F test in a multiple regression tests the null hypothesis that:
    (a) All of the true coefficients are equal to zero.
    (b) All of the true coefficients are not equal to zero.
    (c) Some of the true coefficients are equal to zero.
    (d) Only one true coefficient is equal to zero.

11. If five independent variables are used to predict Y and if there are 30 sets of points in the sample, then the degrees of freedom for variation due to regression would be:
    (a) 4   (d) 25
    (b) 5   (e) 28
    (c) 24

12. For the previous problem, the degrees of freedom for variation due to residuals, i.e., due to deviations from regression, would be:
    - (a) 4
    - (b) 5
    - (c) 24
    - (d) 25
    - (e) 28

13. The table F value for the overall F test at the .05 level of significance would be:
    - (a) 2.7763
    - (b) 2.7587
    - (c) 2.6207
    - (d) 2.6030
    - (e) 3.1548

14. The maximum number of independent variables that can be used in a regression model:
    - (a) Depends entirely upon the computer package that will be used.
    - (b) Is equal to the sample size, n.
    - (c) Is unlimited.
    - (d) Is n-2.
    - (e) None of the above.

15. If there are more independent variables in a model than there are observations (i.e., sets of data points) then we have a(an) _____ problem.
    - (a) Over specification
    - (b) Multicollinearity
    - (c) Cause and effect
    - (d) Extrapolation

16. A systematic procedure for selecting X variables into the prediction equation from best variable to least explanative is called a(an) _____ regression.
    - (a) Hypothesized
    - (b) Simultaneous
    - (c) Stepwise
    - (d) Overall

17. Linear dependency that may exist among two or more of the independent variables in a regression model is called:
    - (a) A multicollinearity.
    - (b) Simultaneity.
    - (c) A prediction error.
    - (d) A counter-insurgency.

18. A dummy variable is:
    (a) A stupid variable.
    (b) One which is included in model for no specific reason.
    (c) One whose values are either zero or one.
    (d) A variable that cannot tell you much.

19. The existence of a multicollinearity problem could be detected from observing _____ on an SPSS printout.
    (a) A TOLERANCE value near 1.0
    (b) A TOLERANCE value near 0
    (c) A PARTIAL correlation near -1.0 or +1.0
    (d) A PARTIAL correlation near zero

20. Once a regression model has been adequately estimated, then we can correctly infer that by changing the values of the X variables, we can <u>cause</u> the Y value to change.
    (a) True.
    (b) False.

CHAPTER 13: Answers to Review Test

Question	Answer	Text Section Reference
1	c	Introduction
2	b	13.1
3	d	13.1
4	d	13.2
5	c	13.2 equation 13.9
6	d	13.3 and 13.8
7	b	13.4
8	b	13.4
9	b	13.2 and 13.5
10	a	13.5
11	b	13.5
12	c	13.5
13	c	13.5
14	d	13.5
15	a	13.8
16	c	13.6
17	a	13.3 and 13.8
18	c	13.7
19	b	13.7 and 13.8
20	b	13.8

CHAPTER 13: Review Problems

1. A growing concern for the amount of energy being consumed by households is sweeping the country. Suppose you are consulted by a government agency interested in describing the characteristics that explain the amount of electricity used by an individual household. A random sample of houses would be taken from the state in which you live. What independent variables would you suggest using for explaining the variation in the number of kilowatts of electricity used by several homes? Discuss how you would measure these variables in the survey.

2. This problem concerns an experiment that is motivated by the carnival side show where the expert prognosticator will guess your weight within x pounds or give you a prize. Because of physiological differences between males and females, we will consider each sex separately.

    In order to explain the variation in the weights of college males and females, a random sample was taken from commerce students at a major university. Knowing that there are many variables with potential for explaining differences in weights and realizing that the useful variables could be different for the two sexes, only three variables were considered for each sex in this experiment. For males, the height, waist and collar size were measured in inches. For females, the height, waist and dress size were recorded.

    A multiple regression model was used to determine the effectiveness of these three variables for explaining body weight. Each sex was analyzed separately. A modified summary of the results obtained using a computer package is given in Table 13.1 for males and Table 13.2 for females.

    (a) Use the methods of Chapter 8 to construct a 95% confidence interval for the true mean weight of all males in this population.
    (b) Give the pairwise correlations between height and weight and between waist and weight for both males and females. What information do these variables convey?
    (c) State the multiple regression model for each sex and the hypotheses that would be tested by the overall F-test.
    (d) For each sex, determine the value of F, the rejection region and the conclusion that should be reached at the .05 level of significance.

(e) What is the numerical value of R^2?
(f) What is the estimated prediction equation for each sex?
(g) How accurate can you expect the predictions to be? Which equation seems to be better? (Hint: Use R^2 and $s_{Y|123}$).
(h) Explain the meaning of the B and BETA values for the waists of males.
(i) Predict your own body weight using the appropriate prediction equation. What other variables could make the prediction equations better?

3. Determine the relationship between the gasoline mileage (number miles per gallon) and the horsepower of the engine, the weight of the car and the octane rating of the gasoline. Twenty midsize cars are randomly selected and driven under similar conditions by drivers of equal experience.

Table 13.3 contains the raw data for the 20 automobiles and a summary of the computer results for stepwise regression.

(a) Which variable by itself explains the most variation of the gasoline mileages? Why? How much variation does it explain?
(b) Why is weight the second variable to be selected?
(c) Octane is the last variable to be considered for inclusion in the model. Test the hypothesis,

H_o: $B_3 = 0$.

H_a: $B_3 \neq 0$.

to determine if octane should be in the prediction equation. Use alpha of .05.
(d) What prediction equation would you recommend using? Give the standard error of Y and the R^2 for this equation.
(e) Construct a 90% confidence interval on the true coefficient of HORSEPOWER, B_1, using the step at which the stepwise procedure should be stopped.
(f) Suppose an individual left 300 pounds of unnecessary items in the trunk of his car, items such as golf clubs, tools, books, etc. Estimate how much this could affect the gasoline mileage?
(g) Predict the gasoline mileage for a limousine weighing 6000 pounds with 490 horsepower and using 100 octane gasoline.

TABLE 13.1

MULTIPLE REGRESSION FOR WEIGHTS OF COLLEGE MALES

VARIABLE	MEAN	STANDARD DEV	CASES
WEIGHT	174.0926	26.3691	54
HEIGHT	71.5185	2.4320	54
COLLAR	15.6389	.6828	54
WAIST	33.5741	2.2704	54

CORRELATION COEFFICIENTS

	WEIGHT	HEIGHT	COLLAR	WAIST
WEIGHT	1.00000	.57883	.80725	.83741
HEIGHT	.57883	1.00000	.54098	.39612
COLLAR	.80725	.54098	1.00000	.68396
WAIST	.83741	.39612	.68396	1.00000

DEPENDENT VARIABLE..   WEIGHT

ANALYSIS OF VARIANCE	DF	SUM OF SQUARES	MEAN SQUARE	F
REGRESSION	3.	30507.47266	10169.15755	?
RESIDUAL	50.	6345.06438	126.90129	

MULTIPLE R	. ?
R SQUARE	. ?
ADJUSTED R SQUARE	.81750
STANDARD ERROR	11.26505

------------------ VARIABLES IN THE EQUATION ------------------

VARIABLE	B	BETA	STD ERROR B	F
WAIST	.6122253+001	.52713	.93507	42.868
HEIGHT	.1967594+001	.18147	.75719	6.752
COLLAR	.1346063+002	.34854	3.39459	15.724
(CONSTANT)	-.3826850+003			

ALL VARIABLES ARE IN THE EQUATION

TABLE 13.2

MULTIPLE REGRESSION FOR WEIGHTS OF COLLEGE FEMALES

VARIABLE	MEAN	STANDARD DEV	CASES
WEIGHT	124.2500	11.4047	16
HEIGHT	65.7500	2.2061	16
DRESS	8.8750	1.8212	16
WAIST	24.8125	1.4245	16

CORRELATION COEFFICIENTS

	WEIGHT	HEIGHT	DRESS	WAIST
WEIGHT	1.00000	.19343	.81368	.74583
HEIGHT	.19343	1.00000	-.07467	.21745
DRESS	.81368	-.07467	1.00000	.45293
WAIST	.74583	.21745	.45293	1.00000

DEPENDENT VARIABLE..   WEIGHT

ANALYSIS OF VARIANCE	DF	SUM OF SQUARES	MEAN SQUARE	F
REGRESSION	3.	1679.92613	559.97538	?
RESIDUAL	12.	271.07387	22.58949	

MULTIPLE R   . ?
R SQUARE   . ?
ADJUSTED R SQUARE   .82632
STANDARD ERROR   4.75284

---------------- VARIABLES IN THE EQUATION ------------------

VARIABLE	B	BETA	STD ERROR B	F
WAIST	.3427092+001	.42806	1.00735	11.574
HEIGHT	.7623110+000	.14746	.58154	1.718
DRESS	.3950310+001	.63081	.77123	26.236
(CONSTANT)	-.4596566+002			

ALL VARIABLES ARE IN THE EQUATION

TABLE 13.3. REGRESSION ON GAS MILEAGE

CASE	HORSE-POWER	WEIGHT	OCTANE	GASOLINE MILEAGE
1	180.	2800.	85.	20.0
2	195.	3000.	85.	18.9
3	160.	2700.	80.	19.1
4	235.	3200.	95.	16.5
5	175.	2700.	90.	21.3
6	285.	3400.	95.	14.9
7	210.	3000.	90.	17.2
8	230.	3300.	90.	15.2
9	235.	3500.	90.	13.0
10	195.	2800.	85.	18.8
11	290.	3500.	95.	14.0
12	150.	2500.	80.	18.8
13	100.	2100.	80.	22.0
14	290.	3400.	95.	13.3
15	255.	3400.	90.	15.2
16	220.	2900.	90.	17.1
17	280.	2200.	85.	14.6
18	280.	3200.	90.	14.6
19	140.	2200.	90.	21.3
20	145.	3200.	85.	19.0

SUMMARY

VARIABLE	MEAN	STANDARD DEV	CASES
OCTANE	88.2500	4.9404	20
HORSEPOW	212.5000	57.0203	20
WEIGHT	2950.0000	444.2617	20
MILEAGE	17.2400	2.8072	20

CORRELATION COEFFICIENTS

	OCTANE	HORSEPOW	WEIGHT	MILEAGE
OCTANE	1.00000	.72164	.66543	-.60188
HORSEPOW	.72164	1.00000	.64304	-.90751
WEIGHT	.66543	.64304	1.00000	-.71026
MILEAGE	-.60188	-.90751	-.71026	1.00000

REGRESSION ON GAS MILEAGE

DEPENDENT VARIABLE..   MILEAGE

VARIABLE(S) ENTERED ON STEP NUMBER 1..   HORSEPOW

MULTIPLE R   .90751
R SQUARE   .82357   ANALYSIS OF VARIANCE   DF   SUM OF SQUARES   MEAN SQUARE   F
ADJUSTED R SQUARE   .81377   REGRESSION   1.   123.31202   123.31202   84.02552
STANDARD ERROR   1.21143   RESIDUAL   18.   26.41598   1.46755

------ VARIABLES IN THE EQUATION ------   ------ VARIABLES NOT IN THE EQUATION ------

VARIABLE   B   BETA   STD ERROR B   F   VARIABLE   BETA IN   PARTIAL   TOLERANCE   F

HORSEPOW   -.4467827-001   -.90751   .00487   84.026   OCTANE   -.11062   -.18232   .47924   .585
(CONSTANT) -.2673413+002   WEIGHT   -.21602   -.39386   .58650   3.121

* *

VARIABLE(S) ENTERED ON STEP NUMBER 2..   WEIGHT

MULTIPLE R   .92246
R SQUARE   .85094   ANALYSIS OF VARIANCE   DF   SUM OF SQUARES   MEAN SQUARE   F
ADJUSTED R SQUARE   .83341   REGRESSION   2.   127.40977   63.70488   48.52459
STANDARD ERROR   1.14579   RESIDUAL   17.   22.31823   1.31284

------ VARIABLES IN THE EQUATION ------   ------ VARIABLES NOT IN THE EQUATION ------

VARIABLE   B   BETA   STD ERROR B   F   VARIABLE   BETA IN   PARTIAL   TOLERANCE   F

HORSEPOW   -.3783960-001   -.76860   .00602   39.515   OCTANE   .23536   .39039   .41008   2.877
WEIGHT   -.1364972-002   -.21602   .00077   3.121
(CONSTANT) -.2930758+002

* *

VARIABLE(S) ENTERED ON STEP NUMBER 3..   OCTANE

MULTIPLE R   .93470
R SQUARE   .87366   ANALYSIS OF VARIANCE   DF   SUM OF SQUARES   MEAN SQUARE   F
ADJUSTED R SQUARE   .84997   REGRESSION   3.   130.81113   43.60371   36.88028
STANDARD ERROR   1.08734   RESIDUAL   16.   18.91687   1.18230

------ VARIABLES IN THE EQUATION ------   ------ VARIABLES NOT IN THE EQUATION ------

VARIABLE   B   BETA   STD ERROR B   F   VARIABLE   BETA IN   PARTIAL   TOLERANCE   F

HORSEPOW   -.4364292-001   -.88648   .00666   42.958
WEIGHT   -.1875651-002   -.29684   .00079   5.600
OCTANE   .1337361+000   .23536   .07885   2.877
(CONSTANT)   .2024508+002

MAXIMUM STEP REACHED

CHAPTER 13: Solutions to Review Problems

1. Solution (Partial)

   Appliances which are big energy users should be included as separate independent variables and could be measured as dummy variables - one, if the household has the appliance, zero otherwise. Some of these variables would be air conditioning, resistance heating, electric stove, electric hot water heater, separate food freezer, and heat pump. Other variables concerning the physical composition of the house that could be important are square footage, number of floors, amount of insulation, typical thermostat settings for winter and summer, number of windows, and use of storm windows. Variables concerning family members could be the number of people living in the house, the age of household head and the income.

2. Solution

   a. The mean male weight is $\bar{X} = 174.09$ with $s = 26.3691$ for $n = 54$. For 95% confidence and 53 degrees of freedom, $t_{(.025,53)}$ is about 2.00.

   $$\bar{X} \pm t_{(.025,53)} \cdot s/\sqrt{n} = 174.09 \pm 2.0 \cdot 26.3691/\sqrt{54}$$

   $$= 174.09 \pm 7.18$$

   I am 95% confident that the true mean weight for all males in the population falls in the interval, 166.91 and 181.27.

   b.
Correlations, r, for	Height & Weight	Waist & Weight
Males	.57883	.83741
Females	.19343	.74583

   For both sexes waist has a stronger linear relation with weight, indicating that if <u>simple</u> linear regressions were calculated, waist would be a better prediction of weight than height.

   c. The models are as follows:

   For males,

   $$Y = A + B_1 \cdot \text{HEIGHT} + B_2 \cdot \text{COLLAR} + B_3 \cdot \text{WAIST}$$

For females,

$$Y = A + B_1 \cdot \text{HEIGHT} + B_2 \cdot \text{DRESS} + B_3 \cdot \text{WAIST}$$

(Y is weight in pounds).

For both cases, the overall F tests:

H_o: $B_1 = B_2 = B_3 = 0$
H_a: at least one is not zero.

d. See equation 13.15. For males, the calculated F is

$$F = \frac{10169.15755}{126.90129} = 80.1343$$

and the Table F is $F_{(.05,3,50)} = 2.8$ (approximately). For females, the calculated F is

$$F = \frac{559.97538}{22.58949} = 24.7892$$

and the Table F is $F_{(.05,3,12)} = 3.4903$. For both sexes, reject H_o. We conclude that at least one of these three variables is useful for predicting weight.

e. See equation 13.9 and Table 13.1 in textbook. For males,

$$R^2 = \frac{\text{SS due to Regression}}{\text{SS Total}} = \frac{30507.}{30507. + 6345.}$$

$$= .828$$

For females

$$R^2 = \frac{1679.9}{1679.9 + 271.1} = .861$$

f. The prediction equations are:

MALE: $\hat{Y} = -382.69 + 6.122 \cdot \text{WAIST} + 1.968 \cdot \text{HEIGHT} + 13.46 \cdot \text{COLLAR}$

FEMALE: $\hat{Y} = -45.97 + 3.427 \cdot \text{WAIST} + .7623 \cdot \text{HEIGHT} + 3.950 \cdot \text{DRESS}$

g. For males, $s^2_{Y|123} = 126.90$ and $s_{Y|123} = 11.27$ while $R^2 = .828$. For females $s^2_{Y|123} = 22.589$ and $s_{Y|123} = 4.75$ while $R^2 = .861$. A little higher proportion of the variation is explained for females, and variance (and standard error) of Y around the regression plane is much smaller for females.

h. For male waists, B = 6.122 means that for each additional one inch in waist size, the predicted weight will be 6.122 pounds higher; BETA = .527 means that for each standard deviation increase in waist (2.27 inches), the weight would increase .527 standard deviations (26.3691) or 13.90 pounds.

i. ? How close?

3. Solution

a. Horsepower, since is entered into the equation first using the stepwise procedure; i.e., if three individual simple regressions were done, horsepower would produce the largest $r^2 = .82$.

b. Weight, because of the two variables not in the equation, has the largest absolute partial correlation with Y adjusted for horsepower; or because it has the largest F value, 3.121 compared to .585.

c. H_o: $B_3 = 0$. (octane)

H_a: $B_3 \neq 0$.

We are testing the importance of the octane variable given that horsepower and weight are in the equation. The test statistic is $F = 2.877$ which is compared to a table F value based on 1 (because we are testing only one coefficient) and 16 degrees of freedom; i.e., $F_{(.05,1,16)} = 4.4940$. Since the computed $F = 2.877$ is smaller than the table F value, we do not reject H_o.

We conclude that there is insufficient information to justify the usefulness of octane rating for predicting gasoline mileage; hence, we will leave it out of the prediction equation.

d. The prediction equation from step 2 is used:

$$\hat{Y} = 29.308 - .03784 \cdot \text{HORSEPOWER} - .001365 \cdot \text{WEIGHT}.$$

$R^2 = .85094$ and $s_{Y|12} = 1.146$.

e. $b_1 = -.03784$ and $s_{b_1} = .00602$.

For 90% confidence and 17 degrees of freedom (always use degrees of freedom for RESIDUAL), $t_{(.05,17)} = 1.740$.

Therefore, $b_1 \pm t_{(.05,17)} \cdot s_{b_1}$ is

$-.03784 \pm 1.740 \cdot .00602$.

I am 90% confident that the true coefficient, B_1, of the variable horsepower falls in the interval $-.04831$ to $-.02737$.

f. For weight, $b_2 = -.001365$. An additional 300 pounds would decrease the mileage by about 0.4 miles per gallon.

g. Using the equation for step 2, $\hat{Y} = 2.576$ miles per gallon. However, the equations are valid for only midsize cars with values of the variables within the ranges used to determine the equation. This is a problem of extrapolation and the predicted value is not reliable.

Time Series Analysis

Major Topics and Key Concepts

14.1 The Notion of a Time Series

14.2 Components of Time Series

- Trend.
- Seasonal variations.
- Cyclical variations.
- Irregular variations.

14.3 Smoothing

- Moving averages; example 1.
- Exponential smoothing; equation 14.2 and example 2.

14.4 Measuring the Components of a Time Series

- Multiplicative model: value = $T \cdot S \cdot C \cdot I$.

14.5 Trend Analysis

- Use regression analysis as in Chapters 12 and 13.
- Use polynomial line for curved trend; example 2.

14.6 Seasonal Indices

- Numbers which vary from a base of 100.
- Two methods of calculating.

14.7 Finding Seasonal Measures Using Multiple Regression

- Use of dummy variables.
- Based on Additive Model instead of Multiplicative Model.

14.8 Cyclical Variation

- Periods longer than a year.
- Ratio of moving average to trend for subannual data.

- Ratio of actual value to trend for annual data.

14.9 Using Time Series Results and Forecasting

- Adjustments for seasonal variation; equation 14.15.
- Trend equation and seasonal indices.
- Forecasting succeeding periods from present data and seasonal indices.

CHAPTER 14: Review Test

1. The sequence of values recorded over time for any piece of data forms what is called:
   (a) A long-term trend.
   (b) An index.
   (c) A time series.
   (d) A chronological regression trend line.

2. Time series analysis consists of breaking time series into component parts so that concrete statements can be made about them.
   (a) True.
   (b) False.

3. Trends in time series:
   (a) Are characterized by steady variable rates of change.
   (b) Are characterized by only slightly variable rates of change.
   (c) Can be represented by straight lines or smooth curves.
   (d) (a) and (c) only.
   (e) All of the above.

4. Seasonal variations can be found in data which are recorded:
   (a) Biennially.
   (b) Monthly.
   (c) Quarterly.
   (d) Both (b) and (c).
   (e) All of the above.

5. Cyclical variations are movements in a time series that are:
   (a) Recurring.
   (b) Longer than one year.
   (c) Constant in amplitude and duration.
   (d) Both (a) and (b).
   (e) All of the above.

6. Irregular variations are movements in a time series which:
   (a) Cannot be predicted using historical data.
   (b) Are caused by general economic conditions.
   (c) Are periodic in nature.
   (d) All of the above.

7. Large variations in time series can be smoothed by using a number of smoothing techniques. Two such techniques are:
   (a) Regression Trend Lines and Averaging.
   (b) Moving Averages and Seasonal Devariation.
   (c) Exponential Smoothing and Moving Averages.
   (d) The Least Squares Line and the Linear Trend Line.

8. In the moving average technique, the average is said to be moving in the sense that when the average for each new month is calculated, a new value of the original time series is brought into the calculation of the average and one is dropped out.
   (a) True.
   (b) False.

9. A seven-month moving average:
   (a) Loses 3 months at the beginning and 4 months at the end of a time series.
   (b) Loses 6 months at the beginning of the time series.
   (c) Loses 3 months only at the beginning of the time series.
   (d) Loses 3 months on each end of the time series.

10. Smoothed values always contain more variation than the original values.
    (a) True.
    (b) False.

11. If the statistician wants a smoothed series that has most of the random, volatile variation taken out of it, he should:
    (a) Use a large smoothing constant, α.
    (b) Use a smoothing constant with a value of .5.
    (c) Use a small smoothing constant.
    (d) Use a smoothing constant with a value of 1.0.

12. The values in a time series are usually expressed in terms of the four components by using:
    (a) A multiplicative model.
    (b) An additive model.
    (c) A probability model.
    (d) A normal distribution model.

13. Trends in time series data are most commonly represented by:

    (a) Exponentially smoothed lines.
    (b) Curved lines.
    (c) Curvilinear lines.
    (d) Least squares lines.

14. A seasonal index value of 100 for a particular month indicates that the month has:

    (a) Maximum seasonal variation.
    (b) No seasonal variation.
    (c) 100% more variation than the base month.
    (d) 100 units of seasonal variation.

15. The most appropriate method of obtaining seasonal indices for time series which have strong trends and cyclical variations is:

    (a) Comparing each season's actual value to a yearly moving average.
    (b) Comparing the seasonal mean value to a grand mean value.
    (c) To use a table of indices published for such circumstances.
    (d) A combination of exponential smoothing and moving averages.

16. When a multiple regression analysis with dummy variables is used as a means of describing seasonal variation instead of the classical approach, the primary difference is that the additive model of regression assumes that seasonal variations produce the same amount of change while the multiplicative model assumes the same percentages of change.

    (a) True.
    (b) False.

17. If the trend value is divided into the moving average, the resulting ratio is:

    (a) A measure of irregular variations.
    (b) A ratio of long-term trend to seasonal variation.
    (c) A measure of the cyclical component.
    (d) None of the above.

18. One of the most common uses of seasonal indices is adjustment of actual data to account for seasonal variation.

    (a) True.
    (b) False.

19. A deseasonalized figure:
    (a) Is free of irregular variation.
    (b) Is obtained by dividing the actual figure by an appropriate seasonal index expressed as a percentage.
    (c) Shows trends and cyclical variations more clearly.
    (d) All of the above.

20. Forecasting next period's sales by using only this period's data and seasonal indices is a method that:
    (a) Can be used for long-range forecasts.
    (b) Can be used for short- and long-range forecasts.
    (c) Should be avoided.
    (d) Can only be used to obtain short-range forecasts.

CHAPTER 14: Answers to Review Test

Question	Answer	Text Section Reference
1	c	14.1
2	a	14.1
3	e	14.2
4	d	14.2
5	d	14.2
6	a	14.2
7	c	14.3
8	a	14.3
9	d	14.3
10	b	14.3
11	c	14.4
12	a	14.4
13	d	14.5
14	b	14.6
15	a	14.6
16	a	14.7
17	c	14.8
18	a	14.9
19	c	14.9
20	d	14.9

CHAPTER 14: Review Problems

1. Sales of the Elledge and Griffin Company, 1969-1979, are shown below in thousands of dollars:

Year	Sales	Year	Sales
1969	450	1974	550
1970	490	1975	560
1971	530	1976	555
1972	510	1977	525
1973	540	1978	580
		1979	600

   (a) Find the three-year moving average for the series.
   (b) Find the exponentially smoothed series using $\alpha = .4$.

2. Net quarterly sales of Gahin Sales and Service Company were as follows during 1975-1980 inclusive (thousands of dollars):

Quarter	1975	1976	1977	1978	1979	1980
1st	120	116	80	88	92	96
2nd	109	105	71	76	77	87
3rd	132	127	87	97	101	103
4th	195	167	116	122	125	130

   (a) Determine the quarterly seasonal indices using the method described in Example 1 of Section 14.6.
   (b) Determine the quarterly seasonal indices using the method described in Example 2 of Section 14.6.

3. If the actual net sales of Gahin in the first quarter of this year were $105,748, what is the seasonally adjusted net sales for the quarter? What would be the expected annual net sales total for the Year? [Use seasonal indices from Problem 2, part (b)].

4. Consider a time series whose first value was recorded in July 1963. The last period for which there are records is October of 1972. How many full months of data are available?

5. Plains States Telephone and Telegraph Company forecasts the demand for its services quite far into the future. This is done so that the necessary financing, hiring, and equipment ordering can be done well in advance of the demand for the service. One of the things that Plains States is interested in is the number of new

telephone installations they will be called upon to make each year. Data indicating the number of installations (in hundreds) made in each of the past 15 years are given below:

Year	Installations	Year	Installations	Year	Installations
1966	15.00	1971	24.00	1976	40.50
1967	14.90	1972	25.10	1977	44.50
1968	15.08	1973	32.71	1978	52.82
1969	17.00	1974	36.11	1979	61.77
1970	20.02	1975	37.55	1980	75.22

(a) What type of trend (linear or curved) might best be fit to this time series?
(b) Find the equation of the least squares linear trend line which fits the time series in Problem 2. Let $t = 1$ be the origin year of the series (1966).
(c) What would the trend line value be for 1984?

6. The following are seasonal indexes for the consumption of electricity in the United States: January, 120; February, 114; March, 107; April, 101; May, 94; June, 91; July, 90; August, 90; September, 92; October, 94; and November, 99. What is the December seasonal index? Assume that the actual consumption for January 1982 turned out to be 11,000 billions of kilowatts. Make a forecast for the total consumption of electricity in 1982.

7. Pacific Finance, a sales finance company, projects quarterly seasonal indexes of 98.1, 123.8, 101.6 and 76.5, respectively, for next year. The company forecasts a $48 million volume of installment financing in the coming year. Use the indices to break down the annual forecast into quarterly forecasts for next year.

CHAPTER 14: Solutions to Review Problems

1. **Solution**

    a. The general formula for finding a three-year average is $\hat{Y}_t = Y_{t-1} + Y_t + Y_{t+1}$ and for 1970 this is $\frac{450 + 490 + 530}{3} = 490$

Year	Sales	3-Year Moving Average	Exponentially Smoothed $\alpha = .4$
1969	450	--	450.0
1970	490	490	466.0
1971	530	510	491.6
1972	510	526.67	499.0
1973	540	533.33	515.4
1974	550	550	529.2
1975	560	555	541.5
1976	555	546.67	546.9
1977	525	553.33	538.2
1978	580	568.33	554.9
1979	600	--	572.9

    b. The general formula for exponential smoothing is: $S_t = \alpha Y_t + (1-\alpha) S_{t-1}$ $\quad 0 \leq \alpha \leq 1$

    Let $S_1 = Y_1$. The calculations for the first three years are as follows:

    $S_1 = Y_1 \hspace{5em} = 450$

    $S_2 = .4(490) + (.6)450 = 196 + 270 = 466$

    $S_3 = .4(530) + (.6)466 = 212 + 279.6 = 491.6$

    The other figures are shown in the table in part (a).

2. **Solution**

    a. The solution can be found by arranging the information in tabular format and obtaining the necessary totals and means.

Year	1st Quarter	2nd Quarter	3rd Quarter	4th Quarter
1975	120	109	132	195
1976	116	105	127	167
1977	80	71	87	116
1978	88	76	97	122
1979	92	77	101	125
1980	96	87	103	130
Quarter Totals	592	525	647	855
Grand Total				2619
Quarter Mean*	98.67	87.50	107.83	142.50
Grand Mean**				109.125
Quarterly Index***	90.42	80.18	98.81	130.58

*Quarter Total/6.

**2619/25 or Σ (Qtr. Means)/4.

***(Qtr. Mean/Grand Mean) x 100.

b. The first steps of the second approach are given in the table on the next page.

Column	1	2	3	4
Year-Qtr.	Actual Data	4-Qtr. Moving Average of Col. 1	2-Qtr. Moving Average of Col. 2	Actual Data as % of Mvg. Average
1-1	120	--	--	--
1-2	109	--	--	--
1-3	132	139.00	138.500	95.31
1-4	195	138.00	137.500	141.82
2-1	116	137.00	136.375	85.06
2-2	105	135.75	132.250	79.40
2-3	127	128.75	124.250	102.21
2-4	167	119.75	115.500	144.59
3-1	80	111.25	106.250	75.29
3-2	71	101.25	94.875	74.84
3-3	87	88.50	89.500	97.21
3-4	116	90.50	91.125	127.30
4-1	88	91.75	93.000	94.62
4-2	76	94.25	95.000	80.00
4-3	97	95.75	96.250	100.78
4-4	122	96.75	96.875	125.94
5-1	92	97.00	97.500	94.36
5-2	77	98.00	98.375	78.27
5-3	101	98.75	99.250	101.76
5-4	125	99.75	101.000	123.76
6-1	96	102.25	102.500	93.66
6-2	87	102.75	103.375	84.16
6-3	103	104.00	--	--
6-4	130	--	--	--

QUARTER (from Col. 4)

Year	1st	2nd	3rd	4th
1975	--	--	95.31	141.82
1976	85.06	79.40	102.21	144.59
1977	75.29	74.83	97.21	127.29
1978	94.62	80.00	100.78	125.93
1979	94.36	78.27	101.76	123.76
1980	93.66	84.16	--	--
Mean percentage	88.60	79.33	99.45	132.68
Total of percentages				400.06
Seasonal Indices	88.58	79.32	99.44	132.66
Total				400.00

3. Solution

   The First Quarter seasonal index from part (b) is 88.58.

   The first quarter sales figure of $105,748 is equivalent to an adjusted sales value of $105,748/.8858 = $119,381.35.

   To get the expected annual net sales, we multiply the adjusted first quarter sales by 4.

   $119,381.35 x 4 = $477,525.40

4. Solution

   Answer - 112 months.

5. Solution

   a. A plot shows the points curved.

b. Trend line.

Year	t to X	Installations (Y)	X^2	Y^2	XY
1966	1	15.00	1	225.000	15.00
1967	2	14.90	4	222.010	29.80
1968	3	15.08	9	227.4064	45.24
⋮	⋮	⋮	⋮	⋮	⋮
1980	15	75.22	225	5658.0484	1128.30
	$\Sigma X = 120$	$\Sigma Y = 512.28$	$\Sigma X^2 = 1240$	$\Sigma Y^2 = 22238.1392$	$\Sigma XY = 5199.25$

$\overline{X} = 8$ $\overline{Y} = 34.152$

$\Sigma x^2 = 280$, $\Sigma y^2 = 4742.7526$, $\Sigma xy = 1101.01$

Therefore,

$$b = \frac{\Sigma xy}{\Sigma x^2} = 3.9322; \quad a = \overline{y} - b\overline{x} = 2.6945$$

Hence,

$\hat{Y} = 2.6945 + 3.9322\, t$ where t = (year − 1965)

c. For 1984, t = 19; Y = 77.406 hundred installations.

6. <u>Solution</u>

   December seasonal index = 108.

   The January index of 120 implys January consumption is (1.2 · AVG MONTHLY CONSUMPTION). Hence, the average monthly consumption is 11,000/1.2 = 9166.66. Therefore, the forecasted annual consumption is 12 · 9166.66 = 110,000 billion kilowatts.

7. <u>Solution</u>

   Average quarterly volume is 48/4 = 12 million. For each quarter,
   1st: 12 · .981 = 11.762 million dollars
   2nd: 12 · 1.238 = 14.856 million dollars
   3rd: 12 · 1.016 = 12.192 million dollars
   4th: 12 · .765 = 9.18 million dollars

Index Numbers

Major Topics and Key Concepts

15.1 The Simple Aggregate Index

- Base year.
- Common units.
- General Formula.

15.2 Price Indices With Quantity Weights

- Weighted aggregate indices.
- Laspeyres index numbers.
- Paasche index number.
- Fisher's "ideal" index number.
- Measuring physical volume changes.

15.3 Selecting a Base Year

- A fairly recent period.
- A period of normal activity.
- Comparable.
- Availability of data.

15.4 The Consumer Price Index

- "Old CPI".
- CPI-U: Consumer Price Index for All Urban Consumers.
- CPI-W: Revised Consumer Price Index for Urban Wage Earners and Clerical Workers.
- WPI: Wholesale Price Index.

15.5 Uses of Indices

- "Constant dollars" and "purchasing power."
- GNP: Gross National Product.
- Dow-Jones Industrial Average.

15.6 Summary

CHAPTER 15: Review Test

1. Index numbers are used to measure the variation of a time series from a base value selected at some fixed point in time.

    (a) True.
    (b) False.

2. An index number is:

    (a) A standard of comparison used mainly by unions in collective bargaining.
    (b) Seasonally adjusted.
    (c) A ratio, multiplied by 100 to put it in the form of a percentage.
    (d) All of the above.

3. A simple aggregate index is found by:

    (a) Dividing the base year aggregate total by any comparison year.
    (b) Dividing aggregate yearly totals by the aggregate base year total.
    (c) Step (a) multiplied by 100.
    (d) Step (b) multiplied by 100.

4. By changing _____, a statistician could construct a simple aggregate index to express almost any percentage price change he desired.

    (a) The base year
    (b) Quantities purchased
    (c) Units of measure
    (d) City where sampling is done

5. When an index number is constructed to reflect the combined effect of changes in a number of varying quantities, the simple aggregate index:

    (a) Should be used.
    (b) Should not be used.
    (c) May be appropriate.
    (d) Would most accurately reflect the changes.

6. A simple aggregate price index can be used legitimately only:

    (a) When all prices are expressed in the same units.
    (b) When equal amounts of each item are purchased.
    (c) When prices are relatively stable.
    (d) Both (a) and (b) are correct.
    (e) All of the above are correct.

7. The weighted aggregate indices for expenditures on a market basket of items would reflect changes from year to year in:

    (a) Prices only.
    (b) Amount purchased.
    (c) Both prices and amount purchased.
    (d) None of the above.

8. Laspeyres index numbers:

    (a) Measure the change in cost for a fixed buying pattern.
    (b) Use base year quantities as weights for the prices.
    (c) Are more convenient to use on a continuing basis than the Paasche index numbers.
    (d) All of the above.

9. In order to construct Laspeyres index numbers, we must use the base-year prices as weights for the quantities.

    (a) True.
    (b) False.

10. When using _____ index numbers, a new set of quantity weights is needed for each period.

    (a) Laspeyres
    (b) Paasche
    (c) Fisher's "ideal"
    (d) All of the above
    (e) Only (b) and (c) are correct.

11. Fisher's "ideal" index number:

    (a) Has many "ideal" mathematical properties.
    (b) Is the easiest index number to calculate.
    (c) Is becoming more and more popular because of its practical applications.
    (d) Is found by squaring the product of the Laspeyres and Paasche index numbers.

12. Index numbers can be used to measure physical volume changes.

    (a) True.
    (b) False.

13. When selecting a base period for an index number series, which of the following criteria should be considered?
    (a) The base period should be fairly recent.
    (b) It should be a period of ideal economic activity.
    (c) It should be a period in which unemployment rates are low.
    (d) It should not be a census year because of the excess data which would not be available for other years in the series.
    (e) All of the above.

14. The base year commonly being used today is 1967. It was selected because:
    (a) It is a fairly recent year.
    (b) It was the last year of relatively stable prices prior to the price surges in 1968 and the early 1970s.
    (c) Of its availability of data.
    (d) All of the above.

15. One limitation of the CPI is that it does not reflect changes in the quality of a product.
    (a) True.
    (b) False.

16. The price index which measures the changes in prices from 1967 of goods and services that were typically purchased in the 1972-73 period by <u>all</u> urban consumers is the
    (a) Old CPI.
    (b) CPI-U.
    (c) CPI-W.
    (d) WPI.

17. Which of the following does <u>not</u> represent a difference between the CPI-U and the old CPI?
    (a) More urban areas are sampled for the CPI-U.
    (b) The weight for clothing is lower for the CPI-U.
    (c) The weights for consumer expenditures were determined more recent years for the CPI-U.
    (d) The base has been updated for the CPI-U.
    (e) The percent of the population covered is at least doubled for the CPI-U.

18. An important use of the Wholesale Price Index, WPI, is

    (a) Adjusting dollar values to "constant dollars."
    (b) To express the purchasing power of the dollar.
    (c) In forecasting rising or falling retail prices.
    (d) All of the above.
    (e) Only (a) and (c) are true.

19. The reciprocal of the _____ is used to express the "purchasing power" of the dollar.

    (a) GNP
    (b) WPI
    (c) CPI
    (d) Dow-Jones Industrial Average

20. The Dow-Jones Industrial Average is found by dividing the closing prices of the stocks by the number of stocks represented.

    (a) True.
    (b) False.

21. The most well-known economic series in the United States is

    (a) The Gross National Product, GNP.
    (b) CPI.
    (c) WPI.
    (d) New York Stock Exchange Index.

22. One reason that the New York Stock Exchange Index was developed was due to the lack of representativeness of the Dow-Jones Industrial Average.

    (a) True.
    (b) False.

CHAPTER 15: Answers to Review Test

Question	Answer	Text Section Reference
1	a	15.1
2	c	15.1
3	d	15.1
4	c	15.1
5	b	15.1
6	d	15.1
7	c	15.2
8	d	15.2
9	b	15.2
10	e	15.2
11	a	15.2
12	a	15.2
13	a	15.3
14	d	15.3
15	a	15.4
16	b	15.4
17	d	15.4
18	e	15.4
19	c	15.5
20	b	15.5
21	a	15.5
22	a	15.5

CHAPTER 15: Review Problems

1. The following data were obtained from the records of the Scooda Produce Company:

Commodity and Unit	Quantities Sold (000's)			Prices Charged		
	1978	1979	1980	1978	1979	1980
Potatoes/lb.	12	13	15	2.40	2.50	2.60
Milk /gallon	44	48	54	1.40	1.60	1.75
Apples /bushel	20	22	24	5.40	6.15	6.50
Eggs /dozen	80	88	97	.80	.95	1.15

(a) Find the simple aggregate index of prices for each year using 1978 as the base year.
(b) Find the weighted aggregate index number for each year (1978 base year).
(c) Find the Laspeyres retail price index for each year using 1979 as the base year.

2. Assume that the price indices in part (b), Problem 1, reflect consumer prices in a given community. A wage earner's salary was $9,500 in 1978, $11,500 in 1979, and $14,500 in 1980. Compare his real income for the three years. Did his buying power increase or decrease?

3. Consider the following data on price and production of the three basic softwoods in the United States.

	1975		1980	
	Price (per bd. foot)	Production (millions of bd. ft.)	Price	Quantity
Douglas Fir	.20	800	.26	790
Southern Pine	.30	830	.34	800
Western Pine	.22	720	.27	750

(a) Find the weighted aggregate index number.
(b) Find a physical volume index for 1980 weighting quantities with 1975 prices.

4. A certain resort owner is interested in calculating an "Index of Fun" for his resort. He is interested in showing that the cost of a week's fun at his resort has not increased as much as the consumer price index and is thus a good bargain in these days of inflation. His

index consists of three items in the average guest's bill: room, food, and recreation. The average prices for these items and the quantities used by the average guest in one week are given below for 1970 and 1980. Calculate the Index of Fun for this resort (use Laspeyres method). Assume that the CPI is 160. Will the resort owner be justified in advertising that his prices have increased less than the general price level in the economy?

	1970		1980	
Room	$35.00/day	6 days	$55.00/day	5 days
Food	$ 4.50/meal	15 meals	$ 7.50/meal	15 meals
Rec.	$ 4.00/event	20 events	$ 5.00/event	34 events

5. Suppose that in 1972 the CPI was 122.3 and that in 1973 it was 128.5. If a laborer in 1972 earned $2.95 per hour and in 1973 earned $3.20 per hour, did his purchasing power increase or decrease? By how much?

6. In conducting wage negotiations for your union at the beginning of this year, you have access to the following information:

	Six years ago	Five years ago	Four years ago	Three years ago	Two years ago	Last year
Mean Weekly Take-Home Pay	$109.50	$112.20	$116.40	$125.08	$135.40	$138.10
Consumer Price Index	112.8	118.2	127.4	138.2	143.5	149.8

In which year did the employees have the greatest buying power in terms of real weekly wages? What percent increase in present wages is required (if any) to provide the same buying power that the employees enjoyed in the year in which they had the highest real wages?

7. Suppose the buying power of the dollar has decreased 25 percent over the past five years. During this same period, a given state employee has had a 20 percent raise in pay. What percent increase in pay would provide the same buying power which the employee had at the beginning of the period?

8. Assume that a firm's plant was built in 1950 at a cost of $1,250,000. Assume that at that time the construction cost index was 150.0. The estimated life of the building was 50 years with no salvage value. Depreciation was taken on a straight-line basis ($25,000 per year). In fiscal 1979, the firm reported a net income of $64,000 after depreciation had been allowed on the plant. Compute the net income which would have been reported if depreciation could be taken on replacement cost, assuming that the construction cost index for 1979 was 390.0.

CHAPTER 15: Solutions to Review Problems

1. **Solution**

   a. The simple aggregate price index is obtained by adding the prices for each year and then dividing the yearly totals by the base year total. The result is multiplied by 100.

	1978	1979	1980
Totals	$10.00	$11.20	$12.00
Index	100	112	120

   b. The weighted aggregate indices are found by Formula (15.2) in the text

   1978: $\frac{262.40}{262.40} \times 100 = 100$

   1979: $\frac{328.20}{262.40} \times 100 = 125.1$

   1980: $\frac{401.05}{262.40} \times 100 = 152.8$

   Note that the weighted indices show much greater rises than the simple aggregate index. This indicates that the largest price increases occurred in the items with large quantity sales (eggs and milk).

   c. The Laspeyres index number is found by multiplying the prices by the base year quantities and then proceeding as if we were calculating a simple aggregate index. (See Formula 15.3)

Item	1979 q	1978 $p_{78} q_{79}$	1979 $p_{79} q_{79}$	1980 $p_{80} q_{79}$
Potatoes	13 lbs.	31.20	32.50	33.80
Milk	48 gal.	67.20	76.80	84.00
Apples	22 bu.	118.80	135.30	143.00
Eggs	88 doz.	70.40	83.60	101.20
Total		287.60	328.20	362.00
Price Index		87.6	100.0	110.3

Note that these indices are not comparable to those found in parts (a) and (c) since these have a different base year.

2. Solution

To determine his constant dollar wages, divide his salary by the year's corresponding CPI, expressed as variation from 1.0 instead of from 100.

$$\frac{\$9,500}{1.00} = \$9,500$$

$$\frac{\$11,500}{1.251} = \$9,192.65$$

$$\frac{\$14,500}{1.528} = \$9,489.53$$

Thus, we see that his buying power decreased when compared to his 1978 constant dollar wages. Although his buying power in 1980 increased over what it had been in 1979, it was still slightly below the 1978 standard. Thus his earnings have just about kept pace with inflation. This example illustrates the frustration a wage earner can feel when he or she experiences a healthy increase in wages but sees no change in standard of living.

3. Solution

a. First find the weighted total for each year:

$$\Sigma P_{75} \cdot Q_{75} = .20(800) + .30(830) + .22(720)$$
$$= 567.4$$

$$\Sigma P_{80} \cdot Q_{80} = .26(790) + .34(800) + .27(750)$$
$$= 679.9$$

Weighted aggregate index is

$$\frac{\Sigma P_{80} \, Q_{80}}{\Sigma P_{75} \, Q_{75}} \times 100 = 119.8$$

b. The physical volume index holds price fixed and measures change in quantities.

$$\Sigma P_{75} \cdot Q_{80} = .20(790) + .30(800) + .22(750)$$
$$= 563.0$$

$$\Sigma P_{75} \cdot Q_{75} = 567.4$$

Using equation 15.6,

$$I_{80:75} = \frac{\Sigma P_{75} \cdot Q_{80}}{\Sigma P_{75} \cdot Q_{75}} \times 100 = 99.2$$

4. **Solution**

   For Laspeyres index use equation 15.3.

   $$\Sigma P_{80} \cdot Q_{70} = 55(6) + 7.50(15) + 5.00(20) = 542.5$$
   $$\Sigma P_{70} \cdot Q_{70} = 35(6) + 4.50(15) + 4.00(20) = 357.5$$

   Therefore, $L_{80:70} = \frac{542.5}{357.5} = 151.7$

   The resort owner would be justified in advertising his claim.

5. **Solution**

Year	CPI	Wage	Constant-Dollar Wage
1972	122.3	2.95	2.95/1.223 = 2.41
1973	128.5	3.20	3.20/1.285 = 2.49

   In terms of constant dollar wage, his purchasing power increased by $.08.

6. **Solution**

   For each year, determine the constant-dollar wage by PAY/(CPI/100):

Six Years ago	Five Years ago	Four Years ago	Three Years ago	Two Years ago	Last Year
97.07	94.92	91.37	90.51	94.36	92.19

   Hence, the greatest buying power was six years ago. To determine the change in wage needed to regain the same buying power:

   $$92.19(1+R) = 97.07$$

   Thus, $R = .0529$ or a 5.29% raise is needed.

7. Solution

A 25% decrease in buying power implies the employee can buy only 75% of what she could have bought 5 years ago.

Purchasing Power = (1/CPI) x 100; hence, CPI = 100/.75 = 133.33. With a 20% increase in pay, index 120, she would still need a raise of

$$120(1+R) = 133.33$$

or R = .1111 or 11.11%.

8. Solution

In constant dollars, the cost of the plant in 1950 is $1,250,000/1.5 = $833,333.33. Building the plant in 1979 would cost $833,333.33 · 3.9 = $3,250,000. Straight-line depreciation for 50 years with no salvage value would allow a $65,000 depreciation in 1979. The firm's 1979 income before depreciation is $64,000 + 25,000 = $89,000. Had replacement cost been allowed as a basis for depreciation, the net income would have been $89,000 - 65,000 = $24,000!

Decision Theory

Major Topics and Key Concepts

16.1 The Payoff Table

- A rectangular array showing the economic consequences for a combination of several alternative actions that could be chosen and "states of nature" that could occur.
- Examples.
- Dominant actions.

16.2 Prior Probabilities

- Incorporating the likelihood of the "states of nature" occurring into the payoff table.
- Exact probabilities.
- Subjective probabilities.

16.3 Expected Monetary Value

- Sum of the products of the payoffs and the probabilities of the payoff occurring.
- Calculated for each alternative action.

16.4 The Opportunity Loss Table

- Construction.
- Expected opportunity loss.

16.5 Expected Value of Perfect Information (EVPI)

- Expected cost of uncertainty—maximum amount worth spending for perfect information.
- Expected payoff under certainty—the average payoff per play.

16.6 Utility

16.7 Decision Criteria

- Maximax or optimist's criterion.
- Maximin or pessimist's criterion.
- Maximum likelihood criterion.
- Equal likelihood criterion.
- Expected monetary value or Bayes' criterion.

16.8 Decision Trees

- A graph of a payoff table.
- Sensitivity analysis.

CHAPTER 16:   Review Test

1. The "best" course of action in decision problems depends on:
   (a) Economic consequences.
   (b) Alternative actions.
   (c) Probabilities of occurrence.
   (d) All of the above.

2. A payoff table is a rectangular array of numbers in which:
   (a) The columns usually represent alternative actions.
   (b) The rows usually represent alternative actions.
   (c) The columns usually represent events.
   (d) The numbers in rows and columns represent probabilities.

3. A negative value in a profit payoff table indicates that
   (a) A mistake has been made.
   (b) The action or event should be ignored in the decision-making process.
   (c) There is to be a payoff.
   (d) An opportunity loss table would be more appropriate.

4. Payoff tables can be used in problems where all the consequences are costs--in which case, costs should be:
   (a) Entered as negative payoffs (payouts).
   (b) Entered as positive figures.
   (c) Expressed as percentages.
   (d) Entered in reverse order.

5. Payoff tables usually indicate which course of action should be taken.
   (a) True.
   (b) False.

6. An action is said to _____ other actions when its payoffs are better than the others regardless of what event occurs.
   (a) Nullify.
   (b) Supersede.
   (c) Cancel.
   (d) Dominate.

7. Subjective probabilities in decision making:
   (a) Are meaningless.
   (b) May be legitimate and can be entered into probability columns of payoff tables.
   (c) Are mere guesses and are usually far from correct.
   (d) Should never be used.

8. Probabilities entered into a payoff table before any experimentation or sampling is done to evaluate them are called _____.
   (a) Subjective probabilities.
   (b) Expected probabilities.
   (c) Prior probabilities.
   (d) Primal probabilities.

9. The combination of possible action payoffs with event probabilities results in a figure called:
   (a) EMV
   (b) EOL
   (c) EVPI
   (d) EPUC

10. In one-time decision problems, expected monetary value should be viewed as an artificial number which shows the weighted average of the payoffs associated with an action.
    (a) True.
    (b) False.

11. In general, the opportunity loss for any action in a particular row of a payoff table is computed:
    (a) By taking the reciprocal of the associated payoff.
    (b) By dividing it into the best payoff in that row.
    (c) As the sum of that action's payoff and the least payoff in that row.
    (d) As the difference between that action's payoff and the best payoff in that row.

12. The opportunity loss table:
    (a) Contains all positive or zero values.
    (b) Contains at least one zero value in each row of the table.
    (c) Measures the cost of uncertainty.
    (d) All of the above.

13. It is <u>always</u> true that the action with the lowest expected opportunity loss will also be the action with the best expected monetary value.

    (a) True.
    (b) False.

14. The maximum amount a decision maker would pay for a perfect predictor of events is called the expected value of perfect information (EVPI) and is equal to:

    (a) EMV.
    (b) EPUC + EMV.
    (c) EOL for the best action.
    (d) None of the above.

15. The average payoff per event using a perfect predictor is called the:

    (a) Expected payoff under certainty.
    (b) Expected monetary value.
    (c) Event value of perfect information.
    (d) Sure payoff.

16. Monetary value serves as a good measure of utility when the payoffs involved are large relative to the overall wealth, assets, or budget of the decision maker.

    (a) True.
    (b) False.

17. The _____ criterion ignores event probabilities and <u>all payoffs</u> except the best one.

    (a) Maximax.
    (b) Maximin.
    (c) Maximum likelihood.
    (d) Equal likelihood.

18. The pessimist is likely to use the _____ criterion.

    (a) Maximum likelihood.
    (b) Bayes'.
    (c) Maximax.
    (d) Maximin.

19. When choosing a course of action involves a great deal of preparation for carrying out the action even before the event is known, the _____ criterion may be most appropriate.

    (a) Bayes'.   (c) Maximum likelihood.
    (b) Maximax.   (d) Maximin.

20. Most decision makers would be well off if they used only the criterion of expected monetary value in making their decisions.

    (a) True.
    (b) False.

21. A decision tree

    (a) Is a graphical aid to decision making.
    (b) Can be made for any payoff table.
    (c) Can be useful when a problem calls for a sequence of decisions which must be made.
    (d) All of the above.

22. A decision tree is always evaluated

    (a) From left to right.
    (b) From right to left.
    (c) From bottom to top.
    (d) From top to bottom.

23. Points in the tree where decisions or choices must be made are usually represented by _____ and are called _____.

    (a) Circles; decision forks.
    (b) Triangles; decision points.
    (c) Rectangles; decision forks.
    (d) Rectangles; decision nodes.

24. In evaluating decision trees, the expected value on one branch cannot be used in the calculation of expected values further down the tree.

    (a) True.
    (b) False.

25. When the time period covered by a decision tree is a year or more, the monetary values in the problem should be:

    (a) Subjected to sensitivity analysis.
    (b) Replaced by more recent figures.
    (c) Discounted.
    (d) Used with extreme care.

CHAPTER 16: Answers to Review Test

Question	Answer	Text Section Reference
1	d	16.1
2	a	16.1
3	c	16.1
4	b	16.1
5	b	16.1
6	d	16.1
7	b	16.2
8	c	16.2
9	a	16.3
10	a	16.3
11	d	16.4
12	d	16.4
13	a	16.4
14	c	16.5
15	a	16.5
16	b	16.6
17	a	16.7
18	d	16.7
19	c	16.7
20	a	16.7
21	d	16.8
22	b	16.8
23	d	16.8
24	b	16.8
25	c	16.8

CHAPTER 16: Review Problems

1. Dr. Fred Morgan is considering construction of a condominium complex as an investment at South Padre Island, Texas. He has sufficient property to build five units. In order to determine how many units to construct, he has set up the following payoff table. The number of units rented is a monthly mean. The payoffs are in thousands of dollars.

Events \ Actions	Probability of renting	Build 0	Build 1	Build 2	Build 3	Build 4	Build 5
0 units rented	.05	0	-8	-11	-14	-17	-20
1 unit rented	.10	0	7	4	1	-2	-5
2 units rented	.20	0	7	19	16	13	10
3 units rented	.30	0	7	19	31	28	25
4 units rented	.25	0	7	19	31	43	40
5 units rented	.10	0	7	19	31	43	55

   (a) Find the action with the best EMV.
   (b) Find the maximax decision.
   (c) Find the maximin decision.
   (d) Find the maximum likelihood decision.
   (e) Find the equal likelihood decision.

2. In Problem 1, what is the maximum amount Dr. Morgan should be willing to pay for perfect information as to which event will occur?

3. Mr. J. B. Highroller owns several metal fabricating firms in the Midwest. He has been quite successful in recent years and is now considering a major expansion. He plans to expand by purchasing 100 percent of the stock in either Company A or Company B.

   Company A has 1 million shares of stock outstanding and they could be acquired for $6 million. In calculating the desirability of the purchase, Mr. Highroller would charge the purchase price of the stock against the stock's first three year's earnings. The annual earnings per share of Company A are currently $2.50 per share. However, the future earnings depend on anti-pollution legislation of the state where Company A is located. If Bill No. 1 is passed, J. B.'s accountants figure that Company A's earnings per share will drop to $2.25 for the next three years. If Bill No. 2 is passed, the per share earnings are figured at $2.00 for three years. Mr. Highroller's executive vice president has investigated the chances of passage for these bills and says they are:

   No legislation passed: 20%

   Bill No. 1: 30% chance of passage

   Bill No. 2: 50% chance of passage.

Company B can be purchased for $5 million. Its earnings per share are currently $3, and there are 800,000 shares outstanding. The future earnings of the company depend on the outcome of a suit which has been filed against it by two competing firms. If Company B loses the court fight, it will have to pay damages and fines that are expected to reduce earnings per share to $1 for the next few years. Since Mr. Highroller calculates the desirability of his investments based on results in the first three years, the amount of time it will take to settle the suit affects his evaluation. J.B.'s lawyer estimates that there is a .4 probability that the suit will be ruled on one year from the purchase date, and a .6 probability that it will be ruled on two years after the purchase. He also estimates that there is a .5 probability of Company B's winning or losing the suit regardless of the date of its settlement.

Draw a decision tree for Mr. Highroller's problem. Determine the expected monetary values for each purchase based on the first three years' results. (Note that you should deal in total cash flows, not per-share flows.) Which company should he purchase? Ignore discounting.

4. Consider the following payoff table. Entries in the table indicate thousands of dollars. Find the EMV of each action. Which is the best action to take?

Events \ Actions	Probabilities	A_1	A_2	A_3
E_1	.36	8	25	-10
E_2	.21	15	-18	2
E_3	.43	-5	10	7

5. For Problem 4, construct the corresponding opportunity loss table. Find the EOL for each action.

6. For Problems 4 and 5, what is the EVPI?

7. In Problems 4-6, what is EPUC?

8. For Problem 4, find each of the following:
   (a) The maximax decision.
   (b) The maximin decision.
   (c) The maximum likelihood decision.
   (d) The equal likelihood decision.

9. Consider the following payoff table. The figures in the table are project costs in thousands of dollars.

Events \ Actions	Probabilities	A_1	A_2	A_3	A_4
E_1	.25	25	20	18	25
E_2	.45	20	30	35	25
E_3	.30	15	10	12	9

   (a) Are any of the actions dominated? If so, eliminate any dominated actions. Find the EMV of the best remaining action.
   (b) What is the maximax decision?
   (c) What is the maximin decision?
   (d) What is the maximum likelihood decision?
   (e) What is the equal likelihood decision?
   (f) What is the maximum amount a decision maker faced with this problem would be willing to pay for information about which event is going to occur?

10. Consider the following payoff table. The figures are profits in thousands of dollars.

Events \ Actions	Probabilities	A_1	A_2	A_3
E_1	.60	90	50	40
E_2	.40	60	190	390

(a) Which action has the highest EMV?
(b) How much must the probability on E_1 be before the preferred action changes?
(c) How much can the profit $E_2A_3 = 390$ fall before A_3 is no longer a desirable action (assuming the original probabilities hold)?

CHAPTER 16:   Solutions to Review Problems

1. Solution
   a. By multiplying the probabilities and the actions together and then summing the columns, we obtain the EMV of each action. These are:

      Build 0 = 0; Build 1 = 6.25; Build 2 = 16.00;
      Build 3 = 22.75; Build 4 = 25.00;
      Build 5 = 23.50.

      Thus, binding four units has the highest EMV and would be the action to take under this criterion.

   b. Under the maximax criterion, the decision maker selects that action which has the best payoff associated with it--disregarding probabilities. Since 55 is the highest payoff in the table, building 5 units would be selected.

   c. Under the maximin criterion, the decision maker selects that action which will maximize minimum outcomes. Here, the decision would be to build zero units since its zero payoff is better than the negative payoffs (losses) associated with the other actions: -8, -11, -14, -17, and -20, respectively.

   d. Under the maximum likelihood criterion, we first determine which event is most likely to occur. This is the average rental of three units per month. Now we determine the payoffs associated with this most probable event. We select the action with the best payoff. Here, the building of 3 units has the best payoff, 31.

   e. Under the equal likelihood criterion, equal probabilities are assigned to all events (1/6 = .167). EMV's are then determined as before. Since all event probabilities are equal, we can also arrive at EMV by summing the columns and dividing by the number of possible events. The EMV's are 0, 4.50, 11.50, 16.00, 18.00, 17.50. Thus, the decision to build 4 units would be the best under this criterion.

2. Solution

   First we convert the payoff table in Problem 1 to an opportunity loss table. (See Section 16.5 for procedures.)

Events \ Actions	Probability of renting	Build 0	Build 1	Build 2	Build 3	Build 4	Build 5
0 units	.05	0	8	11	14	17	20
1 unit	.10	7	0	3	6	9	12
2 units	.20	19	12	0	3	6	9
3 units	.30	31	24	12	0	3	6
4 units	.25	43	36	24	12	0	3
5 units	.10	55	48	36	24	12	0
EOL		30.05	23.80	14.05	7.30	5.05	6.55

The <u>lowest</u> expected opportunity <u>loss</u> (5.05) is also the EVPI. Dr. Morgan should spend no more than $5,050 for the information.

This figure could also have been obtained through EPUC. This figure is:

$$\text{EPUC} = (.05)(0) + (.10)(7) + (.20)(19) + (.30)(31)$$
$$+ (.25)(43) + (.10)(55) = 30.05 \ .$$

Then $\text{EVPI} = |\text{EPUC} - \text{EMV}^*|$

$$= |30.05 - 25.00| = 5.05 \ .$$

3. **Solution**

   The decision tree for Mr. Highroller's problem is found on the next page.

4. **Solution**

   For A_1: $\text{EMV}_1 = .36(8) + .21(15) + .43(-5) = 3.88$

   For A_2: $\text{EMV}_2 = .36(25) + .21(-18) + .43(10) = 9.52$

   For A_3: $\text{EMV}_3 = .36(-10) + .21(2) + .43(7) = -.17$

   Hence, action A_2 has the highest EMV and would be the best action under this criterion.

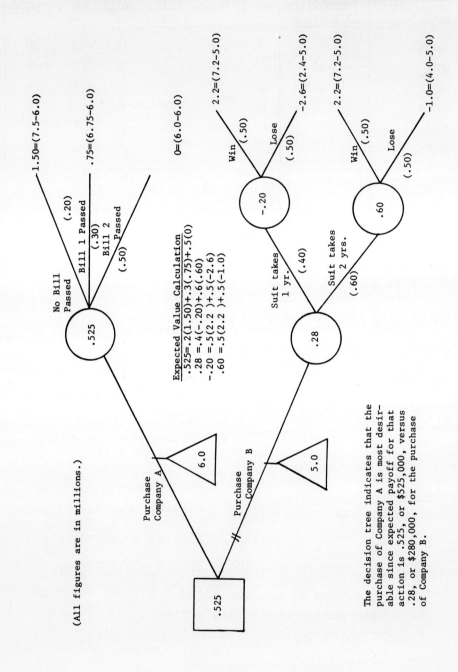

5. Solution

Opportunity Loss Table

	Probabilities	Action 1	Action 2	Action 3
Event 1	.36	17	0	35
Event 2	.21	0	33	13
Event 3	.43	15	0	3

$EOL_1 = 17(.36) + 0(.21) + 15(.43) = 12.57$

$EOL_2 = 0(.36) + 33(.21) + 0(.43) = 6.93$

$EOL_3 = 35(.36) + 13(.21) + 3(.43) = 16.62$

6. Solution

   EVPI = minimum of EOL_i = 6.93 .

   The investor would be willing to pay up to 6.93 thousand dollars to learn in advance which event is going to occur.

7. Solution

   EPUC = Sum of probability times the highest payoff for the event

   = .36(25) + .21(15) + .43(10)

   = 16.45 .

8. Solution

   a. Maximax decision is A_2 since 25 is the largest payoff in the table.

   b. Maximin decision is A_1 since the minimum payoffs for the three actions are -5, -18, -10, and the maximum of these is -5.

   c. Maximum likelihood decision is A_2 since event 3 is the most likely to occur (i.e., it has the highest probability) and the maximum payoff for event 3 is 10.

   d. For equal likelihood, assume the probability for each event occurring is 1/3.

$$EMV_1 = 1/3(8) + 1/3(15) + 1/3(-5) = 6.0$$

$$EMV_2 = 1/3(25) + 1/3(-18) + 1/3(10) = 5.667$$

$$EMV_3 = 1/3(-10) + 1/3(2) + 1/3(7) = -.333$$

Hence, A_1 has the highest expected monetary value.

9. **Solution**

   Note that <u>costs</u> are given and not payoffs.

   a. No actions are dominated. For each action,

   $$EMV_1 = .25(25) + .45(20) + .30(15) = 19.75$$

   $$EMV_2 = .25(20) + .45(30) + .30(10) = 21.50$$

   $$EMV_3 = .25(18) + .45(35) + .30(12) = 23.85$$

   $$EMV_4 = .25(25) + .45(25) + .30(9) = 20.20 .$$

   Hence, the action with the lowest expected cost is A_1.

   b. The maximax decision is A_4 since 9 is the lowest cost.

   c. The maximin decision is either A_1 or A_4. The highest costs are 25, 30, 35, 25; the lowest of these is 25.

   d. The maximum likelihood decision is A_1; 20 is the lowest cost for the event with the highest probability.

   e. For the equal likelihood decision, assume each event has probability of 1/3 of occurring.

   $$EMV_1 = 20, \ EMV_2 = 20, \ EMV_3 = 21.67, \ EMV_4 = 19.67$$

   Action A_4 would have the lowest expected cost.

   f. Opportunity Loss Table (i.e., higher costs)

	Probabilities	A_1	A_2	A_3	A_4
E_1	.25	7	2	0	7
E_2	.45	0	10	15	5
E_3	.30	6	1	3	0

$$EOL_1 = .25(7) + .45(0) + .30(6) = 3.55$$
$$EOL_2 = .25(2) + .45(10) + .30(1) = 5.30$$
$$EOL_3 = .25(0) + .45(15) + .30(3) = 7.65$$
$$EOL_4 = .25(7) + .45(5) + .30(0) = 4.00 \ .$$

The lowest expected opportunity loss is $3,550.

10. <u>Solution</u>

   Note that we are given <u>payoffs</u>.

   a. $EMV_1 = 78$, $EMV_2 = 106$, $EMV_3 = 180$.

   Action A_3 has the higest EMV.

   b. $EMV_1 = 90p + 60(1-p) = 30p + 60$

   $EMV_3 = 40p + 390(1-p) = -350p + 390$

   $EMV_1 = EMV_3$ if $30p + 60 = -350p + 390$

   or $p = .868$.

   c. IF EMV_3 falls below 106, then it is no longer desirable. Let X = the profit to which it can fall.

   $EMV_3 = .6(40) + .4X$

   $106 = 24 + .4X$

   $X = 205$ or $205,000.

   Since the profit is now $390,000, if it falls more than $390,000 - $205,000 = $185,000, then A_3 would no longer be desirable.

17

Decisions and Experiments

Major Topics and Key Concepts

17.1 Subjective Probabilities

- Influencing factors.
- Checking consistency by use of joint probabilities.
- Constructing a cumulative probability curve.

17.2 Revisions of Probabilities

- Use of objective information.
- Conditional probabilities.
- Prior probabilities.
- Combining conditional and prior probabilities.
- Marginal probabilities.
- Revised or posterior probabilities.
- Proportions and the use of the binomial formula.
- Why use subjective probabilities?

17.3 When to Experiment

CHAPTER 17: Review Test

1. Which of the following are elements of decision problems?
   (a) There are always alternatives to choose among.
   (b) The alternatives have certain states of nature.
   (c) There are variable payoffs which are conditional only upon what action is taken.
   (d) All of the above.

2. The problem with probabilities based on past experience is that by using them we automatically assume that the future will be like the past.
   (a) True.
   (b) False.

3. A person making subjective probability estimates has no way of checking for realistic estimates.
   (a) True.
   (b) False.

4. The consistency of subjective probabilities can be checked by:
   (a) Comparison with other subjective probabilities.
   (b) Constructing a cumulative probability distribution.
   (c) Calculating the joint probabilities of compound events.
   (d) Making sure that they add up to a total of 1.5.

5. Given any two points on a probability curve, the area under the curve segment described by those two points can be found by:
   (a) Integral calculus.
   (b) Rectangular approximation.
   (c) Constructing a cumulative probability curve.
   (d) All of the above.

6. The following probabilities are taken from a cumulative probability curve: P(35) = .31; p(50) = .45. What is the probability that an event defined by the range 35-50 will occur?
   (a) .76
   (b) -.14
   (c) .14
   (d) Cannot be determined.

7. It is usually easier to construct a cumulative probability curve and obtain probabilities from it rather than use the rectangular approximation method.

    (a) True.
    (b) False.

8. A formula called Bayes' Rule can be used to combine known _____ and _____ probabilities.

    (a) Joint, marginal.
    (b) Prior, conditional.
    (c) Prior, posterior.
    (d) Joint, revised.

9. The components which are summed to give the marginal probability are called:

    (a) Revised probabilities.
    (b) Posterior probabilities.
    (c) Joint probabilities.
    (d) Conditional probabilities.
    (e) None of the above.

10. The revised probabilities are equal to the _____ divided into _____.

    (a) Prior probabilities, marginal probability.
    (b) Marginal probability, prior probabilities.
    (c) Joint probabilities, marginal probability.
    (d) Marginal probability, joint probabilities.

11. The conditional probabilities are always the probabilities of observing the objective information obtained, given that the event specified in the row is true.

    (a) True.
    (b) False.

12. In using the tabular method for revising prior probabilities, conditional probabilities are found in the same manner for all problems.

    (a) True.
    (b) False.

13. Prior probabilities for a decision problem just add up to _____, but the sum of any values appearing in the event column is _____.

    (a) 1.00; always greater than 1.00.
    (b) 100; equal to or less than 1.00.
    (c) 1.00; meaningless.
    (d) .10; always greater.

14. When events are proportions, then conditionals can be obtained by using _____.
    (a) The normal probability table.
    (b) The binomial formula or tables.
    (c) By converting to percentages.
    (d) t-distribution tables.

15. When the experimental, or objective, information is used to revise a set of prior probabilities, all of the prior values usually change.
    (a) True.
    (b) False.

16. It will usually pay to perform an experiment and revise a payoff table's prior probabilities when:
    (a) The EVPI is low and the cost of experimentation is low.
    (b) The EVPI is high and the cost of experimentation is high.
    (c) The EVPI is high and the cost of experimentation is low.
    (d) The EVPI is low and the cost of experimentation is high.

CHAPTER 17: Answers to Review Test

Question	Answer	Text Section Reference
1	a	17.1
2	a	17.1
3	b	17.1
4	c	17.1
5	d	17.1
6	c	17.1
7	a	17.1
8	b	17.2
9	c	17.2
10	d	17.2
11	a	17.2
12	b	17.2
13	c	17.2
14	b	17.2
15	a	17.2
16	c	17.3

CHAPTER 17: Review Problems

1. Dr. Grant Calder recently purchased a stock option on the Chicago Options Exchange. The option expires next month, when Dr. Calder will be in Europe on his annual vacation. He is not sure if he should sell now or simply let the option be exercised on the expiration date. At expiration an option typically has either positive or negative value based on the price at which the option was purchased and the value of the stock on which the option has been written. Currently, the stock is at a price where Dr. Calder could just break even in exercising his option. An investment newsletter predicts with "80 percent confidence" that the stock will rise by next month. However, Dr. Calder has just learned that the Wall Street Journal plans to run an article on this company during the next week. He has been informed that the article will be "somewhat negative." When Dr. Calder called his broker about selling the option, the broker sent him a copy of a study which the brokerage had recently completed on the relationship between news stories and stock prices. Dr. Calder was particularly interested in a table in that study. The table is presented below:

Movement of Stock Prices in Month after Article	Type of Article		
	Favorable Article	Neutral Article	Unfavorable Article
Higher	50	150	25
Lower	20	180	50

What should Dr. Calder's revised probabilities be for this problem?

2. The Robson Manufacturing Company has three production lines. They were constructed at different times during the company's expansion years of 1958 to 1973. The first line is the oldest, and it produces 20 percent defectives. The second was installed in 1965 and produces 5 percent defectives. The newest line produces only 1 percent defectives. Since the defective rates differ so widely and since the lines cost about the same to operate, the newer lines are operated most of the time, and the oldest line is used only when demand for the product is very high. Last week the old line was used to produce 2,000 units, the 1965 line was used to produce 8,000 units, and the new line was used to

produce 10,000 units. However, last week a batch of units (500 units make up a batch) was sent to a customer, and the batch was painted red. The usual color of the units is white. The Production Manager, Mr. G. Mangum, is trying to determine who is responsible for the apparent prank. First, he would like to know which production line produced the red units.

(a) What are the prior probabilities for each line?
(b) If Mr. Mangum noted that in a sample of 25 red units, four were defective, how would this change his ideas about which line produced the mysterious units?

3. There is a .3 chance that the nation's economy will enter a recession next year. There is a .8 chance that Company XYZ will experience a drop in sales if a recession occurs. What is the probability that the nation will suffer a recession and Company XYZ's sales will drop?

4. A production manager wishes to construct a probability curve which indicates the likelihood of his company's producing various quantities of the company's product in the next quarter. From prior experience, he feels that there is a 50-50 chance that production will reach a level of more or less than 50,000, a 25 percent chance that production will reach a level of more than 60,000, and a 25 percent chance that production will be less than 45,000. The company has never produced more than 75,000 or less than 30,000. Construct a cumulative probability curve for the manager.

(a) Use the curve to find the probability that next quarter's production will exceed 65,000 units.
(b) What is the probability that next quarter's production will be between 47,500 and 65,000 units?
(c) What is the probability of producing under 40,000?

5. National Foods Corporation wants to increase the sales of one of its leading breakfast cereals. A concern about the high cost of packaging causes them to consider the desirability of offering the cereal in an 18 ounce box instead of the 10 ounce box presently being sold. This would permit them to reduce their selling price per pound because of the lower packaging costs. The Marketing department thinks that such a move would have a probability of .80 of increasing their sales and profits. A certain city is chosen as a test market in order to test their subjective feeling. Management realizes that the test results may be misleading. Based on historical records and judgment, they believe that the probability of getting favorable test results when

the product tested is a successful national product is about .85, and that the probability of getting favorable test results when the product tested is a dud on the national market is about .35.

(a) What are the prior probabilities of favorable and unfavorable national sales?
(b) What are the conditional probabilities?
(c) What are the joint probabilities?
(d) What is the revised probabilities of favorable and unfavorable national sales?

CHAPTER 17: Solutions to Review Problems

1. Solution

The states of nature that might exist at the expiration date are: Option Has Positive Value or Option Has Zero or Negative Value. Since Dr. Calder is currently at the break-even point, these states are the same as: Stock Price Moves Higher or Stock Price Falls. The prior probabilities for these states are .80 and .20, according to the investment newsletter. However, Dr. Calder has additional information which must be taken into account in this problem. He knows that a slightly unfavorable article is about to appear in the Wall Street Journal. In order to revise the .80 and .20 priors, we must determine the conditional probabilities--that is, what is the likelihood of having the experimental result (an unfavorable article appear) given that the stock will rise (since a rise would lead to a positive value for the stock option). The other conditional probability is the likelihood of an unfavorable article's appearing on a stock whose price will fall during the next month.

The brokerage firm's study showed 225 firms whose stocks had gone up in price. Of those, 25 had unfavorable articles written about them. Thus P(Unfavorable Article|Stock Will Rise) = 25/225 = .11. The study also showed that 250 stocks had gone down in price after articles had been written about them. Thus P(Unfavorable Article|Stock Price Will Fall) = 50/250 = .20.

These probabilities are combined with the priors in the table below to give the joint probabilities of having: (both a stock with a rising price and an unfavorable press reaction), which is (.80)(.11) = .088, and (both a stock with a falling price and an unfavorable press reaction), which is (.20)(.20) = .040. The sum of the joint probabilities is .088 + .040 = .128, which is the marginal probability. The revised probabilities of the stock's rising or falling are obtained by dividing the joints by the marginal: .088/.128 = .69 and .040/.128 = .31.

States of Nature	Priors	Conditionals	Joints	Revised
Price Rises	.80	.11	.088	.6875
Price Falls	.20	.20	.040	.3125
	1.00		.128	1.0000

Thus the appearance of an unfavorable article about the company makes it look less likely that the stock's price will rise; the investment newsletter's opinion overrides the probabilities here so that the option still looks good to hold.

One could easily argue that the appearance of the _Wall Street Journal_ article should completely void the opinion of the investment newsletter. If this is the case, then we would look only at those companies on which unfavorable articles have been published. There are 75 such companies. Of these, 25 have had their stocks rise in the month after the article's publication, and 50 have had their stocks fall. Thus the probabilities of rising and falling stock price would be 25/75 = .33 and 50/75 = .67, respectively. These probabilities, however, ignore completely the opinion of the newsletter. They indicate that Dr. Calder should sell his option now since it is more likely that the stock price will fall. His decision depends, then, on whether he wants to take the newsletter's opinions into account.

2. Solution

   a. The states of nature in this problem are: Old Line Did It, 1965 Line Did It, and New Line Did It. The prior probabilities are derived from the production output during the last week.

      P(Old Line)   = 2,000/20,000 = .10.
      P(1965 Line) = 8,000/20,000 = .40.
      P(New Line)   = 10,000/20,000 = .50.

   b. Since the production lines can be viewed as binomial processes, the conditional probabilities consist of the probabilities of getting 4 defectives in a sample of 25 items from each line:

      P(4 of 25 defective|old line) =

      $P(r=4 | n=25, p=.20) = .4207 = .2340 = .1867$

      P(4 of 25 defective|1965 line) =

      $P(r=4 | n=25, p=.05) = .9928 - .9658 = .0269$

      P(4 of 25 defective|new line) =

      $P(r=4 | n=25, p=.01) = 1.0000 - .9999 = .0001$

where the numerical probabilities were taken from Table III of the text, the Binomial Probability Table, page 475. These conditional probabilities and the priors are combined to give the joints, marginal, and revised probabilities in the table below.

States of Nature	Priors	Conditionals	Joints	Revised
Old Line Did It	.10	.1867	.01867	.633
1965 Line Did It	.40	.0269	.01076	.365
New Line Did It	.50	.0001	.00005	.002
	1.00		.02948	1.000

Thus it appears that Mr. Mangum should concentrate his investigation on the people who run the old line. He might very well catch them red-handed.

3. **Solution**

From Chapter 4,

$$P(\text{Recession and Drop}) = P(\text{Recession}) \cdot P(\text{Drop}|\text{Recession})$$

$$= .3 \cdot .8 = .24 \ .$$

4. **Solution**

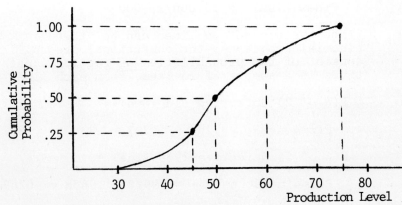

a. About $1.00 - .85 = .15$.

b. About $.85 - .35 = .50$.

c. About .10.

5. <u>Solution</u>
   a. .80 and .20 .
   b. .85 and .35 .
   c. .80 · .85 = .68 and .20 · .35 = .07 .
   d. Using equations 17.1 and 17.2,

   $$P(\text{favorable}) = \frac{.68}{.68 + .07} = .9067$$

   $$P(\text{unfavorable}) = \frac{.07}{.68 + .07} = .0933 .$$

18

Nonparametric Methods

Major Topics and Key Concepts

- These are alternatives to the hypothesis tests of Chapter 9.
- Use them when the original populations from which samples are taken cannot be assumed to be normally distributed.

18.1 The Wilcoxon Test

- An alternative for the one sample t-test and for paired observations.
- Original population should be symmetric.

18.2 The Sign Test

- Also an alternative for the one sample t-test and for paired observations.
- Can be used with any population.
- Not as sensitive as the Wilcoxon Test.

18.3 The Rank Sum Test

- An alternative for the two independent sample t-test.
- Both populations should have the same shape.

18.4 Spearman's Rank Correlation Measure

- An alternative for the sample correlation, r, of Chapter 12.

18.5 A Summary Note on Nonparametric Methods

CHAPTER 18: Review Test

1. A nonparametric test procedure could be used:
    (a) When there are no population parameters to test.
    (b) When the assumptions required for the corresponding parametric test are not satisfied.
    (c) Whenever there are large amounts of data because the methods are computationally easier.
    (d) Anytime the degrees of freedom is large.

2. Nonparametric methods employ the use of ranks and other measures of relative magnitude in place of parameters such as means and variances.
    (a) True.
    (b) False.

3. Nonparametric methods make extensive use of the central limit theorem.
    (a) True.
    (b) False.

4. In computing the Wilcoxon statistic, W, after subtracting the hypothesized median from each observation, the remaining values are:
    (a) Arranged in order of increasing absolute value.
    (b) Arranged in order of increasing magnitude.
    (c) Ranked and assigned algebraic signs in alternating patterns.
    (d) (a) and (c).

5. Given the following observations, $Y_i = X_i - \xi_0$: 3.2, -4.1, 1.5, -.6, 5.8, .33, -7.5, -1.25, and 6.0, using the Wilcoxon method the quantity -4.1 would receive a rank of:
    (a) 2.
    (b) -2.
    (c) 6.
    (d) -5.

6. For observations of _____, the distribution of W can be approximated by use of _____.
    (a) 30 or more, the binomial distribution
    (b) 10 or more, the normal distribution
    (c) 10 or more, the binomial distribution
    (d) 30 or more, the Poisson distribution

7. When doing a Wilcoxon test, if an observed value is equal to ξ_0, it is omitted from the sample, and the sample size is reduced; however, if two or more values

have the same absolute distance from ξ_0, then these ties should be handled by:

(a) Combining the observations.
(b) The subtraction and addition of arbitrary values.
(c) Averaging adjacent ranks.
(d) Assigning opposite signs.

8. The primary assumption necessary for using the Wilcoxon test is symmetry of the population about its median.

(a) True.
(b) False.

9. The sign test is _____ sensitive than the Wilcoxon test and should only be used _____.

(a) More; when the population is symmetric about the mean.
(b) Less; with uniform populations.
(c) More; with a binomial distribution.
(d) Less; when symmetry cannot be assumed.

10. The steps for computing the rank sum statistic include

(a) Combining the two samples into one and underlining the observations from one sample.
(b) Arranging the combined observations in order of decreasing magnitude.
(c) Subtracting the average of underlined observations from the average of all observations.
(d) Ranking the observations according to increasing absolute values.

11. The statistic, V, measures the distance between the sum of ranks for the second sample and sum of the ranks for the first sample.

(a) True.
(b) False.

12. If two samples consist of paired observation, then the most sensitive nonparametric procedure to use is the

(a) Wilcoxon test.
(b) Sign test.
(c) Rank Sum test.
(d) Spearman's Rank Correlation.

13. The normal approximation for the rank test is appropriate if:

(a) $n_1 = n_2$.
(b) Either n_1 or n_2 is greater than 30.
(c) $n_1 + n_2 > 30$.
(d) n_1 and n_2 are both greater than 10.

14. Spearman rank correlation measures:
    (a) The difference between the means of two sets of ranks.
    (b) The variation that exists in two sets of ranks.
    (c) The association between two variables measured with ranks or whose respective values have been replaced by ranks.
    (d) The linear relationship between any two variables.

15. A value of $r_s = -1$ results when:
    (a) The ranks are one unit apart.
    (b) There is no association between the ranks.
    (c) A mistake in calculation is made, since r_s is never negative.
    (d) There is perfect disagreement in the rankings.

16. When using the normal approximation for a nonparametric test, we are really assuming that the original population from which we sample has a normal distribution.
    (a) True.
    (b) False.

CHAPTER 18: Answers to Review Test

Question	Answer	Text Section Reference
1	b	Introduction
2	a	Introduction
3	a	Throughout Chapter
4	a	18.1
5	c	18.1
6	b	18.1
7	c	18.1
8	a	18.1
9	d	18.2
10	a	18.3
11	b	18.3
12	a	18.2
13	d	18.3
14	c	18.4
15	d	18.4
16	b	Throughout Chapter

CHAPTER 18: Review Problems

1. Ten drivers arrested for the third time for driving while under the influence of alcohol were given a safety concern test before and after a two week stay in jail. Does jail affect the safety concern of drinking drivers in this population? A high score shows a concern for safety; the test scores can range from 0 to 100. Use the .12 level of significance. Past experience indicates that changes in safety concern are not normally distributed.

Driver	1	2	3	4	5	6	7	8	9	10
Before Jail	80	72	14	60	60	76	80	55	65	72
After Jail	60	32	1	68	43	68	84	49	76	58

2. A random sample of 16 Phoenix teenagers are asked to watch a new movie and rate the suspense of the movie's ending. A scale of 10 to 50 is used where 10 means no suspense and 50 means wet pants. If the true median suspense score is below 40, the studio will re-make the ending. Use alpha of .05. Past experience indicates ratings of this type are not normally distributed.

Ratings:	44.0	24.8	38.2	40.0
	32.5	26.4	31.0	30.2
	36.0	40.5	34.5	26.6
	36.0	40.0	42.0	49.8

3. Do Problem 2 using the sign test.

4. A new method for making pistons has been proposed. To test whether the new method has increased the compressive strength, twelve sample pistons are made by the new method and compared with ten pistons made by the standard method. The compressive strengths in pounds per square inch are as follows:

   Old: 145 141 146 137 144 135 134 80 138 141
   New: 145 150 148 143 138 145 141 142 146 139 136 140

   The compressive strengths do not follow a normal distribution. Determine if the new method results in stronger pistons. Use the .05 level of significance.

5. A personnel manager and his assistant examined ten job applications and scored them according to the applicants' overall qualifications and potential worth to the firm. Compute the rank correlation coefficient and determine

if it is significant at the 1.0 percent level of significance. High score reflects favorable potential worth.

Applicant	A	B	C	D	E	F	G	H	I	J
Scores by:										
Personnel Manager	72	75	10	20	84	73	80	78	63	68
Assistant	74	75	55	50	80	65	90	85	73	70

CHAPTER 18: Solutions to Review Problems

1. <u>Solution</u>

   We have paired observations; the Wilcoxon test is used. Since n = 10, the normal approximation procedure is valid.

   H_o: $\xi_{BEFORE} = \xi_{AFTER}$ Since $Z_{.06} = 1.555$, the

   H_a: $\xi_{BEFORE} \neq \xi_{AFTER}$ rejection region is

   R: Z < -1.555 or Z > 1.555 .

   (1) Find the difference in scores for each individual, BEFORE-AFTER:

   Y_i: 20 40 13 -8 17 8 -4 6 -11 14

   NOTE: Positive indicates a DECREASE in concern for safety.

   (2) Arranging these in order of absolute value,

   -4 6 -8 8 -11 13 14 17 20 40

   (3) Ranking these values, averaging for ties, and retaining the algebraic sign, we have

   -1 2 -3.5 3.5 -5 6 7 8 9 10

   (4) The sum W = 36. From equation 18.7,

   $$\sigma_W = \sqrt{\frac{n(n+1)(2n+1)}{6}} = \sqrt{\frac{10 \cdot 11 \cdot 21}{6}} = 19.62 \ .$$

   Hence,

   $$Z = \frac{W}{\sigma_W} = 1.835 \ .$$

   Since Z = 1.835 is greater than $Z_{.06} = 1.555$, we reject H_o.

   The stay in jail <u>does</u> affect safety concern. In fact, this study <u>indicates</u> that it decreases the concern for safety.

2. <u>Solution</u>

   This is a one sample test; we will assume the population of ratings is symmetric and use the Wilcoxon test.

$H_o: \xi \geq 40$ Since $Z_{.05} = 1.645$, the

$H_a: \xi < 40$ Rejection region is R: $Z < -1.645$.

(1) Subtract 40 from each observation.

$Y_i = X_i - 40$:   4.0   -15.2   -1.8   0.0
                  -7.5   -13.6   -9.0   -9.8
                  -4.0   0.5   -5.5   -13.4
                  -4.0   0.0   2.0   9.8

(2) Arranging these in order of ABSOLUTE value.

LOW

0   0   0.5   -1.8   2.0   4.0   -4.0   -4.0
-5.5   -7.5   -9.0   9.8   -9.8   -13.4   -13.6   -15.2

                                              HIGH

We delete the two zeros from the sample; now n = 14.

(3) Replacing the above values with their ranks, averaging for ties, and retaining the algebraic signs, we have

x   x   1   -2   3   5   -5   -5
-7   -8   -9   10.5   -10.5   -12   -13   -14

(4) The sum $W = -66.0$. From equation 18.7,

$$\sigma_W = \sqrt{\frac{n(n+1)(2n+1)}{6}} = \sqrt{\frac{14 \cdot 15 \cdot 29}{6}} = 31.859.$$

Therefore,

$$Z = \frac{W}{\sigma_W} = -2.072.$$

Comparing this to the rejection region, we reject H_o and conclude that the ending of the movie should be re-made.

3. <u>Solution</u>

The assumption of symmetry for the population of all ratings for this movie is not needed for the sign test.

The hypothesis and rejection region are the same as for Problem 2.

We count the number of observations greater than 40; r = 4. We delete all values equal to 40; hence, the sample size is n = 14.

Now

$$Z = \frac{r - n \cdot (.5)}{\sqrt{n \cdot (.5) \cdot (.5)}} = \frac{4-7}{\sqrt{3.5}} = -1.60 \, .$$

Since $Z = -1.6$ is not less than $Z_{.05} = -1.645$, we do not reject H_o.

Using the sign test, we would decide not to re-make the ending of the movie.

4. Solution

We have two independent samples; but the populations are not normal; hence, we should use the rank sum test. $H_o: \xi_{NEW} \leq \xi_{OLD}$ versus $H_a: \xi_{NEW} > \xi_{OLD}$

(1) We combine the two samples into one, underlining the observations in the second sample, and arrange them in order of increasing size:

80 134 135 <u>136</u> 137 138 <u>138</u> <u>139</u> <u>140</u> 141 141
<u>141</u> 142 <u>143</u> 144 145 <u>145</u> <u>145</u> 146 <u>146</u> <u>148</u> 150

(2) Now we replace the observations with the ranks, averaging where ties occur:

1 2 3 <u>4</u> 5 6.5 <u>6.5</u> <u>8</u> <u>9</u> 11 11
<u>11</u> 13 <u>14</u> 15 17 <u>17</u> <u>17</u> 19.5 <u>19.5</u> <u>21</u> 22

(3) For Sample 1: Rank Sum = 91, Rank Mean = 9.1

For Sample 2: Rank Sum = 162, Rank Mean = 13.5

V = 13.5 - 9.1 = 4.4 .

From equation 18.12,

$$\sigma_V = \sqrt{\frac{(22)^2 (23)}{12 \cdot 10 \cdot 12}} = 2.780 \, .$$

Therefore, $Z = \frac{V}{\sigma_V} = 1.58$.

The rejection region is R: Z > 1.645.

We conclude that there is insufficient evidence to indicate that the new pistons have increased compressive strength.

5. <u>Solution</u>

First, replace the scores by the ranks for each of the two managers. (We could rank in either direction; we will use rank of 1 as best.)

Applicant	A	B	C	D	E	F	G	H	I	J
Rank by:										
Personnel Manager	6	4	10	9	1	5	2	3	8	7
Assistant	5	4	9	10	3	8	1	2	6	7
D_i	1	0	1	-1	-2	-3	1	1	2	0
D_i^2	1	0	1	1	4	9	1	1	4	0

$\Sigma D_i^2 = 22$.

Hence, $r_s = 1 - \dfrac{6 \Sigma D_i^2}{n(n^2 - 1)} = 1 - \dfrac{6 \cdot 22}{10 \cdot 99} = .8667$.

$H_o: \rho_s = 0$ The Rejection Region is

$H_a: \rho_s \neq 0$ R: $t < -3.355$ or $t > 3.355$

$t_{(8, .005)} = 3.355$

$t = r_s \sqrt{\dfrac{n-2}{1 - r_s^2}} = 4.91$.

Since the calculated $t = 4.91$ is in the rejection region, we reject H_o. There is a positive association between the rankings of the personnel manager and his assistant.